Innovations in Agro Animal Technologies

Innovations in Agro Animal Technologies

Editor-in-Chief
R.B.SRIVASTAVA
and
W.SELVAMURTHY

SATISH SERIAL PUBLISHING HOUSE
403, Express Tower, Commercial Complex, Azadpur, Delhi-110033 (India)
Phone : 011-27672852, Fax : 91-11-27672046
E-mail :info@satishserial.com, hkjain1975@yahoo.com
Website : www.satishserial.com

Published by :

SATISH SERIAL PUBLISHING HOUSE
403, Express Tower, Commercial Complex, Azadpur, Delhi-110033 (INDIA)
Phone : 011-27672852 Fax : 91-11-27672046
E-mail : info@satishserial.com, hkjain1975@yahoo.com

ISBN 978-93-81226-02-5

Typeset at: Harminder *for* Laxmi Art Creation
Printed at: Salasar Imaging Systems, Delhi-110035

डॉ. वी.के. सारस्वत
रक्षा मंत्री के वैज्ञानिक सलाहकार
सचिव एवं महानिदेशक

Dr. V. K. Saraswat
Scientific Advisor to Defence Minister
Secretary Defence R & D and DG DRDO

भारत सरकार
रक्षा मंत्रालय
रक्षा अनुसंधान तथा विकास विभाग
डी आर डी ओ भवन, नई दिल्ली–110 105

Government of India
Ministry of Defence
Department of Defence Research & Development
DRDO Bhawan, New Delhi-110 105

Foreword

The Indian Himalayan Region comprises of Trans, Northwest, West, Central and Eastern Himalayan provinces. The Trans-Himalayan biographic zone is located across the divide of the Greater Himalaya. Cold arid regions of India lie in high altitude ranges of North - Western Himalayas in the northern latitudes of the country. This region represents narrow gorges, rocky cliffs with glistening peaks and scarce population. This zone is characterised by extreme cold, heavy snow fall, low oxygen partial pressure, low humidity and single cropping season resulting in dependence on nearby regions for food and nutrition. The people of these regions encounter various high altitude health problems and food and nutritional insecurity due to reduced food availability especially during winters. Wheat, Barley, Potato, Turnip, Apricot and Apples were the only agricultural and horticultural crops of the region. Due to the peculiar ecology of this region, production potential of agricultural crops and livestock remained unexplored for a long time. Defence Institute of High Altitude Research (DIHAR) earlier known as Field Research Laboratory (FRL), has been endevouring constantly since 1962 under the aegis of Defence Research & Development Organisation (DRDO) towards the development of agriculture, animal sciences and extension to make these difficult terrains sustainable in fresh food availability.

Development of solar green houses, standardization of agricultural and horticultural technologies, research to increase livestock production, development of herbal therapeutics for high altitude maladies, conservation of germplasm and intensive technology transfer are the major research and extension activities of the institute which has considerably alleviated problems of this zone. This edited volume will prove of immense help to the researchers, agriculturist, technocrats, extension scientists and progressive farmers of cold arid region.

This book comprises of chapters contributed by scientists on issues that are important to face the current and future challenges of the cold arid zones for achieving sustainable agriculture and livestock production.

I congratulate the editors of the book Dr. R.B. Srivastava and Dr. W. Selvamurthy for bringing out this publication and all the scientists of DIHAR for making this information available which will go in a long way in keeping scientists and researchers abreast to the advancement in agro-animal technologies for the cold arid zones.

V.K. SARASWAT

Preface

The Trans-Himalayan zone is one of the highest and driest inhabited places on the earth. Climatic conditions of these regions are very harsh such as prevalence of extreme sub zero cold temperature between October to April (- 10 to - 40°C) and scorching heat during July to August (upto + 40°C), low precipitation (9-30 cm) in the form of snow, low relative humidity (20-45%) and high wind velocity (70 nautical mile/hr). These conditions result in sparse vegetation and arctic-like desert conditions, which makes the cold desert an inhospitable terrain for crop cultivation and livestock production. Besides local population, a large number of Army and Para troop personnel are deployed in these regions due to its strategic location. It adds to the huge amount of pressure on the food and nutritional security of this region. The sub-zero temperature prevalent in these regions during the winter months (Oct-April) restrict the crop cultivation to a single season, which is mainly focused on wheat and barley with some turnip, radish, cabbage potatoes, peas and occasionally other vegetables between May to September. The valley is devoid of cultivation during the rest of the year, which necessitates storage of excess produce from the cultivated fields for consumption and demands research initiatives to produce vegetables and fruits during winters.

The high altitude cold desert treeless regions possess extremely hostile environments and therefore men, animals and plants of this region face many biotic and abiotic stresses. Some of the technologies developed by Defence Institute of High Altitude Research (DIHAR) such as multivitamin herbal beverage, herbal tea, herbal antioxidant supplement *etc.* based on the wild thorny bush seabuckthorn have helped locals to earn their livelihood in the cold desert regions and improve their socio-economic condition.

The present book highlights the contributions of various scientists working relentlessly towards development of high altitude agro-animal technologies. The research activities carried out by DIHAR in this region indicate the considerable scope of improvement of the cold deserts. Keeping the above fact in mind, we have endeavored to compile the information on high altitude agro animal technologies along with the research activities of DIHAR for the last 50 years in the form of the book "Innovations in Agro Animal Technologies". In this book, the chapters deal with the various subjects *viz.* vegetable science, seed production technologies for various vegetables production, horticulture, sustainable fruit and vegetable production and its post-harvest management etc. Some chapters describe the research on seabuckthorn and its sustainable utilization, importance of medicinal plant biodiversity, soil health status, DIHAR initiatives for eco-environment improvement in Ladakh, agriculture extension and transfer of various technologies and products developed by DIHAR. The storage and antifreezing studies of biodiesel in cold desert has also been discussed. In animal science, technologies like sustainable fodder production for livestock, strategies for poultry production and hatchability, broiler sheep and goat as a source of meat, dairy technology and Zanskar ponies have been discussed in detail. The book also deals with the physiological response of human beings at high altitude, high altitude maladies and radiation counter measures using bioresources indeginous to high altitude.

The main aim of this book is to present a comprehensive view about the innovations pertaining to high altitude agro animal technologies. The book will make the readers acquainted with the past activities and present research developments of DIHAR and the world. The book owes much to the critical reviewing of manuscripts by Prof. R.S. Chauhan, Prof. R.S. Srivastava, Prof. Ajit Verma and Dr. Brahma Singh. The editors have made atmost efforts to correct errors in chapters. Constructive criticism is welcome which will be thankfully acknowledged.

We extend our sincere gratitude and thanks to Dr. V.K. Saraswat, Scientific Advisor to Raksha Mantri, Secretary Defence Research & Development and DG DRDO and Dr. Ilavazhagan, DLS, DRDO HQ for technical guidance and encouragement. A deep sense of gratitude goes to all the scientists and staff of DIHAR who have authored chapters in this book timely and valuably supported us in bringing this book to a reality.

Leh-Ladakh
Sept., 2011

Editors

Contents

Chapter-1

Innovative Approaches in Vegetable Research to Meet the Nutritional Requisites in Cold Arid Desert Areas

Narendra Singh, Manish Bhoyar, Pal Murugan M., Dorjey Angchok and R.B. Srivastava
Defence Institute of High Altitude Research (DIHAR), DRDO, Leh, Ladakh

1. Introduction

Himalayas, the youngest fold mountains which forms the northern barrier of India has the richest biodiversity amongst the mountain ranges in India. Ladakh is a part of the Indian Himalayas at an altitude of 9000-22,000 feet above mean sea level, and is located between the Greater Himalayas and the Karakoram range. It is also known as the cold arid desert of India, and is characterized by diverse and complex land formations. It lies between 31° 44′57′-32° 59′57′ N latitude and 76° 46′29′-78° 41′ E longitude and is characterized by prolonged sub zero temperature, low annual precipitation (20-30 mm rainfall/snowfall), and low relative humidity. It has many unconquered peaks of impregnable heights, uncharted glaciers and valleys. Soil is sandy, non-fertile with very less water holding capacity, low oxygen content, high UV radiation, very low atmospheric pressure, high wind velocity and rugged terrain with the temperature ranging between +35°C in summer to -35°C during winter. This region is characterized by snow clad mountain peaks and extremely harsh climatic and geographical conditions which poses a challenge for existence of both flora & fauna.

2. Ancient Agriculture Scenario of Ladakh

The harsh climatic conditions prevailing in Ladakh, only favors sparse vegetation and makes the terrain inhospitable for crop cultivation. Some wild plant species however were able to establish themselves in this complex terrain after natural introduction long ago probably due to modifications in their morphological and genetic traits in due course of time for survival and by completing their life cycle in adverse conditions of the region. Majority of such plants are perennial herbs and some of them are suitable for use as vegetables. These were identified by local people over the course of experience gained by hit and trial method. The plants of these species start their growth early and quite a good number of such plants are in full vegetative state much earlier than the introduced cultivated vegetables by the beginning of the summer season (Chaurasia *et al.*, 2000). Before introduction of some conventional vegetable crops such as turnip, radish, carrot, onion, potato and certain leafy vegetables, the people of this region were totally dependent on the wild edible plants species as non-conventional vegetables and this practice can still be seen in far flung villages.

3. Non-conventional Vegetables

The local people of Ladakh utilized the wild edible plants as non-conventional vegetables and used them for culinary purposes. These plants have great nutritional potential, especially during peak winter and early summer season (Bhoyar *et al.*, 2011). Some commonly used wild edible plants are listed below in Table 1.

Table 1 : List of non-conventional vegetables used by local people in Ladakh

S No.	Botanical Name	Local Name	Culinary uses
1.	*Allium humile*	Skotze	Dried leaves are used as condiments by local people.
2.	*Arabidopsis himalaica*	Byi'u la phug	Leaves are used as a green vegetable.
3.	*Arenaria serpyllifolia*	Chiki	The tender leaves are cooked as leafy vegetable.
4.	*Arnebia euchroma*	Demok	The roots are a source of dye, which is used for colouring the food items such as butter, cheese and ghee.

5.	*Capparis spinosa*	Kabra	Tender shoots and young leaves are used as a leafy vegetable after proper boiling. The tender shoots are mixed with curd and prepared a local delicious dish known as 'Tantur'.
6.	*Capsella bursa-pastoris*	Shamso	Tender shoots and young leaves are boiled and cooked as leafy vegetable.
7.	*Cicer microphyllum*	Chana	Pods and seeds are edible.
8.	*Carum carvi*	Kosnyot	Seeds are used as condiments.
9.	*Fagopyrum esculentum*	Dayat	Tender leaves are used as leafy vegetable and seeds are used for making flour to prepare different dishes.
10.	*Lepidium latifolium*	Sangtso	Tender young leaves are used as leafy vegetable.
11.	*Malva verticillata*	Cam-pa ma-ning	Tender shoots and young leaves are boiled and cooked as leafy vegetable.
12.	*Mentha longifolia*	Phololing	Leaves are used for making 'Chatani' and as flavouring agent.
13.	*Orabanche hansii*	Gro-shang-rtse	The tender shoots are used to make soup.
14.	*Oxyria digyna*	Lamanchu	The leaves are very nutritive and used as leafy vegetable and salad.
15.	*Rheum webbianum*	Lacchu	Long and thick leaf stalk used as salad.
16.	*Rhodiola imbricata*	Shrolo	The leaves are used as a leafy vegetable. The tender shoots are mixed with curd and prepared as a local delicious dish known as 'Tantur'.
17.	*Rumex patientia*	Shoma	The fresh leaves are used as a salad and for making sauce.
18.	*Sonchus oleraceus*	Khala	The fresh leaves are used as a salad leafy vegetable.

19.	*Taraxacum sikkimensis*	Khur-mang karpo	The whole plant is used as vegetable and roots are eaten raw.
20.	*Urtica hyperborea*	Zacchout	It is used in famous local dish known as 'Thupka'.

The wild edible plants, collected during the summer months are dried and stored during harsh winter. However, the long term storage attributes to decreased palatability and loss of nutrients.

4. Need for Introduction of New Vegetables

Due to the strategic location, owing to it's borders with Pakistan and China, a large numbers of troops are deployed in Ladakh region. Troops are deployed up to 22000 ft altitude in border regions. Good physical, mental health and high morale is required for the troops in order to defend the borders, especially in harsh regions. Fresh vegetable and fruits play a major role in maintaining physical and mental health of people. As locally produced vegetables cannot meet the huge local demand, fresh vegetables are transported from plains. This region remains landlocked every year due to heavy deposition of snow on the road passes, which necessitates the airlifting of fresh vegetables from Chandigarh.

In spite of heavy expenditure, there is also loss of nutrients due to spoilage of fresh vegetables during transportation. Several Army units are located in extreme hostile environmental conditions of border regions where availability of fresh vegetables is scarce. Due to extreme long winter, the agriculture season is short which extends from May-September in general and in case of particular location its depend upon altitude (Table 2). In spite of multitude of constraints, to ensure year-round availability of fresh vegetables to the troops and civil sector, Defence Institute of High Altitude Research (DIHAR) accepted the challenge of developing appropriate agro technologies which are in-tune with the prevailing socio-economic and environmental conditions of Ladakh region despite the odds of harsh climatic conductions and less productive soil in this cold arid deseart. The developed technologies have been assessed and refined at the farmers' field before further successful demonstration and dissemination at mass scale.

Table 2 : Cropping season and duration at different altitudes

Altitude (m AMSL)	Cropping season	Duration (months)
2400 - 2700	April - October	7.0
2700 - 3000	May - Mid October	5.5
3000 - 3300	Mid May - September	4.5
3300 - 3600	June - Mid September	3.5
3600 - 4000	Mid June - August	2.5

5. Introduction of New Vegetable Varieties and Hybrids

Wide variation in daily and seasonal climate makes it difficult to standardize package of practices for cultivation of vegetable crops in the region. After the establishment of Defence Institute of High Altitude Research (formerly FRL) in 1962 at Leh, systematic experimentation was carried out and some new vegetables were introduced and cultivation practices were standardized. Production techniques of these vegetables were evolved and demonstrated; following which the farmers undertook cultivation of some of the vegetable crops such as cabbage, cauliflower knol-khol, etc. Upto the middle of nineteen nineties, about 40 types of vegetables were introduced and during the last decade more than 30 other types of vegetables have been further evaluated and techniques standardized to obtain optimum yield (Singh *et al.,* 2003). By the end of 2010, DIHAR had introduced 78 different types of vegetables, standardized their cultivation practices and transferred the technology to the local farmers to grow them in their own fields. The availability period of vegetables cultivated in open field during summer season has been enhanced from 2 to 5 months (June-October) by introduction of more number of suitable vegetables/ varieties/hybrids. Techniques have also been standardized to produce more than 20 types of vegetables even during severe winter months (October-April) through greenhouse technology (Singh *et al.,* 2005).

Table 3 : List of introduced new vegetables varieties and hybrids

S. No.	Group	Vegetables	Variety/Hybrids
1.	Cole Crops	Cabbage	Golden Acre, Gonzales (F1), S-25 (F1), Drum Head
2.		Cauliflower	Amazing (F1), Snow crown (F1), Lateman Krishna (F1)
3.		Knol khol	White Vienna, Purple Vienna

4.		Chinese Cabbage	Wong Bok, Optiko
5		Red Cabbage	Red Acre
6.		Broccoli	CBH-1, Packmen, Pirite
7.		Brussels Sprout	Hills Ideal
8.		Kale	Red Russian
9.		Karam Sag	Kashmir
10.		Romanasco	Bejo Sheetal
11.	Root Crops	Carrot	Nantes, Pusa Yamdagini, Samson (F1)
12.		Radish	Pusa Himani, FRL Local Sel, Japanese White
13.		Turnip	Purple Top, White Globe, Pusa Chandrima,
14.		Beet Root	Detroit Dark Red, Action
15.		Swede	FRL selection
16.	Bulb Crops	Onion	Brown Spanish, Sindhu Sweta, Agrifound Dark red
17.		Bunching onion	–
18.		Garlic	Agrifound Parvati, Single Clove
19.		Leek	Suttind Leek
20.	Tuber	Potato	Kufri Chandramukhi, Kufri Jyoti, Kufri Jawahar
21.		Artichoke	Jeruslum, Local
22.	Pulses	Pea	Arkel, Bonneville, Lincoin, VL-3, Pusa Pragati
23.		Snow Pea	(Toledo)
24.		French Bean	Contender, VL Boni-1, Pusa Parvati, French Yellow.
25.		Broad Bean	Sindhu Bramha, Local
26.		Gram	BG-361, BG-374, Pusa-209, C-104
27.		Sem	Kalyanpur Type-2
28.		Cowpea	Pusa Komal
29.		Cluster bean	Pusa Mausami
30.		Moth bean	C-2M-2
31.		Soybean	Bragg

32.	Solanaceous	Tomato	BSS-347 (F1), Tolstoi (F1), Sultan (F1), Sindhu-1
33.		Capsicum	California wonder, Pusa Deepti (F1), and Sindhu Naren.
34.		Brinjal	**Long:** Pusa Hybrid-5, Raveena (Green)
35.		Chilli Peprika	Pusa Jwala, BSS-344(F1)
36.	Salad	Lettuce	Great Lake, Chinese Yellow
37.		Parsley	Masscurled, Pegoda
38.		Celery	Trimmuf
39.		Coriander	Cori-1
40.		Mint	FRL selection
41.		Beet Leaf	Mongol, Sindhu Harit
42.		Spinach	Sporter
43.		Swiss Chard	Red Petiole
44.		Veg. Musturd	ARU-Black, Hill Sarson
45.		Amaranth	Red & Green
46.		Chenopodium	Pusa Red
47.		Fenugreek	Kasuri, Early Bunching
48.		Orich (Ustak)	Sindhu Green
49.		Packchoy	Sindhu Early
50.		Chicory	Local, Pluto
51.		Endive	Bossa
52.		Cress Cresida	–
53.		Portulaca	S-4
54.	Spices	Turmeric	BSR-1
55.		Ginger	Himgiri
56.	Cucurbitaceous	Pumpkin	MPH-1, Sindhu-Yellow, Sindhu Green, Pusa Vishwas
57.		Bottle Guard	Pusa Naveen, Pusa Sandesh, Pusa Hybrid-3, BSS-333
58.		Summer Squash	Australian Green, Pusa Alnkar, Zucchini, Season Opener
59.		Cucumber	Japanese Long green, Pusa Sanyog (F1), Poinsette, Astix
60.		Long melon	Tar green, Long green

61.		Water melon	Asahi Yamato, Sugar Baby, BSS-358, Swapnil, Apoorva
62.		Musk melon	Punjab Hybrid No-1, Madhubala, Hara Madhu, BSS-361
63.		Sarda melon	Sindhu Sweet, Sindhu Leopard, Sindhu Honey
64.		Ridge gourd	Pusa Nasdhar, Harita, BSS-405
65.		Sponge gourd	Pusa Chikini
66.		Bitter gourd	Pusa do Mausmi, Arka Harit, Monsoon Mircle
67.		Round melon	Arka Tinda, JLM-83025
68.		Wax gourd	Mah-1, Co-1
69.		Snap melon	–
70.		Buffalo gourd	FRL Sel
71.	Other crop	Okra	Varsa (F1), Harbhajan
72.		Baby Corn	G-5406
73.	Local Vegetable	Kabra	–
74.		Shantgso	–
75.		Shoma	–
76.		Khala	–
77.		Lamanchu	–
78.		Lachu	–

6. Standardization of Cultivation Practices

Fluctuation in seasonal and diurnal temperatures is very high in cold arid desert region besides harsh climatic condition. Sudden drop in temperature during early stage (April-May) and later stage (September-October) of crop growth is a common phenomenon. This wide variation in climate makes it difficult to standardize common package of practices for vegetables production in this region. To overcome this problem DIHAR has developed crop/location specific practices for optimum productivity over period of time.

6.1 Land Preparation, Manures and Fertilizers

Soil health and fertility play a major role in crop productivity. Soil serve as medium for plant growth, as a nutrient reservoir and as the site for many biological process. The soil analysis conducted by DIHAR has revealed that the type of soil in a particular region of Ladakh depends upon the weathered rock of that location and soils in this region are mostly sandy to sandy loam in nature. It is found that soils of this region are low in organic carbon (0.69%) and phosphorus (9.6 Kg/ha), very low in nitrogen (122kg/ha) and high in Potash (284.3kg/ha) with low water holding capacity (Bhoyar *et al.*, 2009). The vegetables can grow in almost all the cultivable fields in Ladakh. The field should be ploughed once or twice before onset of winter atleast 15 to 20 days before sowing/transplanting of the vegetable crops. The field should be firm and leveled. The doses of fertilizers and organic manure affect the yield of the vegetable crops. Combined use of fertilizers and organic manure gave higher yield. To produce good quality vegetables it is recommended that 2-4 kg (depending on the crop and kind of manure) of well decomposed organic manure should be applied per square meter during last ploughing of the field before sowing/transplanting of the crop. In addition to the above dose, 1-2 kg per square meter well decomposed organic manure can be given after 30-50 days (depending upon the crop) of sowing/transplanting of the crop. Yield may be increased with application of organic manure especially of the crops which required heavy dose nutrients. Soil mixing with night soils is prevalent in Ladakh due to less availability of farm yard manure (FYM).

6.2 Nursery Raising

To get early crop with higher yield, timely plantation is essential especially for the crops which are grown through transplantation. Protected vegetable nursery raising techniques of different vegetables have been standardized to raise the vegetable seedlings for early transplantation in cold arid condition.

Nursery raising is a highly technical aspect and has great impact on planting time, availability period, yield and quality of the produce. Successful production of certain vegetables like solanaceous and cucurbitaceous crops in field conditions depend on early preparation of seedlings. Therefore, to get the early produce of vegetables, practices have been standardized for healthy and vigorous seedlings raising of more than 40 types of vegetable crops in different types of protected conditions like specific polycarbonate greenhouse, trench, polyench and seasonal greenhouse (Singh *et al.*, 2003). DIHAR has developed the capacity to raise more than 12.5 lakh seedlings of different vegetables annually in protected condition. The nursery raised in the greenhouse is ready for transplantation in first week of April (45 days early) for protected vegetable cultivation and this helps to get the produce about 60 days early than open condition with increased yield up to 60 percent depending upon the crop.

6.3 Sowing and Transplanting

The sowing and transplanting of the vegetables can be done in the month of April-May depending on the crop, weather condition and altitude. To facilitate germination, the seeds should be sown at optimum moisture level. A light irrigation should be given just after transplantation of the seedlings. Seed rate and spacing for the sowing the seed or transplanting of the seedlings, days to harvest and yield of vegetable crops are given in Table 4.

Table 4 : Seed rate, Spacing and yield of introduced vegetable crops

S. No.	Vegetables	Seed rate Kg/ha	Spacing		Day to Harvesting (App.)	Yield q/ha (App.)
			Row	Plant		
1.	Cabbage	0.5	60	45	75	450
2.	Cauliflower	0.4	60	45	75	250
3.	Knol khol	1.0	30	30	70	200
4.	Chinese Cabbage	0.5	45	20	70	200
5.	Broccoli	0.5	60	45	75	180
6.	Brussels Sprout	0.6	60	45	90	50
7.	Kale	0.4	30	30	45	250
8.	Karam Sag	0.7	30	20	70	100
9.	Carrot	4.0	30	10	70	250
10.	Radish	10.0	10	10	65	300
11.	Turnip	3.0	30	10	60	350
12.	Beet Root	10.0	30	10	75	250
13.	Swede	3.5	30	10	90	200
14.	Onion	8.0	30	10	120	250
15.	Garlic	800	20	10	150	60
16.	Leek	6.0	15	10	120	100
17.	Potato	2000	50	20	120	350
18.	Artichoke	1600	50	20	120	225
19.	Pea	100	25	05	90	75
20.	Snow Pea	80	30	05	80	125
21.	French Bean	80	30	10	80	80

22.	Broad Bean	100	30	10	90	160
23.	Gram	80	30	15	70	60
24.	Tomato	0.5	60	45	90	550
25.	Capsicum	0.7	60	45	100	60
26.	Brinjal	0.5	60	45	90	100
27.	Chilli	0.5	60	45	90	50
28.	Lettuce	0.8	30	10	45	75
29.	Parsley	0.4	45	15	80	150
30.	Celery	0.25	30	20	80	200
31.	Coriander	30	20	0.5	50	80
32.	Mint	–	15	05	45	90
33.	Beet Leaf	18	30	10	50	120
34.	Spinach	30	30	05	50	80
35.	Swiss Chard	15	30	08	45	150
36.	Veg. Musturd	0.6	30	10	45	70
37.	Amaranth	0.6	30	10	60	60
38.	Chenopodium	0.25	30	15	60	50
39.	Fenugreek	25	30	03	70	100
40.	Orich (Ustak)	04	45	15	70	60
41.	Packchoy	0.8	30	15	50	80
42.	Chicory	01	30	15	60	70
43.	Endive	0.5	45	20	70	75
44.	Cress Cresida	0.6	30	10	75	40
45.	Pumpkin	06	400	200	90	250
46.	Bottle Guard	04	300	150	90	150
47.	Summer Squash	08	150	75	60	300
48.	Cucumber	2.5	150	100	70	80
49.	Long melon	2.5	150	100	70	120
50.	Water melon	2.5	150	100	90	160
51.	Musk melon	2.0	150	100	85	120
52.	Sarda melon	2.0	150	100	90	150
53.	Ridge gourd	2.5	150	100	95	60
54.	Sponge gourd	2.5	200	100	80	75
55.	Bitter gourd	2.5	150	100	90	40
56.	Round melon (Tinda)	2.0	150	100	90	50
57.	Wax gourd	2.5	300	200	110	150
58.	Snap melon (Phoot)	2.0	150	100	80	120
59	Buffalo gourd	2.5	300	150	110	20
60.	Okra	12	45	20	75	40
61.	Baby Corn	60	45	20	80	80

6.4 Irrigation

In cold desert areas, water continues to be the scarce commodity not only for irrigation but even for drinking and other domestic uses. This difficulty has been experienced very frequently, in spite of the fact that important rivers namely Indus, Suru, Zanskar and their tributaries originate from the glaciers of Ladakh. The existing resources are further declining due to heavy biotic pressure and lack of management of existing resources. In most Himalayan cold deserts water is brought in channels from glacial melts for irrigating the fields. Flooding the fields with the glacial water for improving crop productivity is also common. The deposition of fresh silt with unweathered minerals (especially lime) forms glacier source of fresh salts. Low water holding capacity of soil and long photo period necessitated frequent irrigation for the vegetable cultivation in the cold arid region.

6.5 Weeding and Hoeing

The field should be free from weeds. Generally, 2-3 weeding and hoeing are required in all the vegetable crops.

7. Organic Farming

Organic farming is one of the several approaches found to meet the objectives of sustainable agriculture. Many techniques used in organic farming like inter-cropping, mulching and integration of crops and livestock are not alien to various agriculture systems including the traditional agriculture practiced in old countries like India. The chemicals used in crop production such as fertilizers, insecticides are health hazardous besides damaging the soil and environment. Organic farming makes positive contribution for the protection of soil, environment and human health. Non decomposition of animal and plant waste is a serious problem in cold desert.

Adverse effects of modern agricultural practices not only on the farm but also on the health of all living things and thus on the environment have been well documented all over the world. Application of technology, particularly in terms of the use of chemical fertilizers and pesticides all around us has persuaded people to think aloud. Their negative effects on the environment are manifested through soil erosion, water shortages, salinization contamination, genetic erosion, etc. So it is imperative to develop, dessiminate and adopt organic farming practices, such as vermicomposting, mulching, Integrated Nutrient Management *etc.* in Ladakh region due to its fregile nature of ecosystem.

7.1 Vermicomposting

Vermicomposting is the practice of using worms to transform farm waste into a nutrient-rich compost. Worms are efficient food waste digesting machines that eat over half their body weight in organic matter per day. The eating castings that they create are rich in nutrients and completely natural.

Vermicompost application in the Ladakh is very much useful because soil of Ladakh is poor in texture (mostly sandy) and structure, less water holding capacity with low nutrients and organic matter. The subzero temperature prevails in Ladakh prevents decomposition of waste material, when compared to other region because of low microbial activity. The survivability of earthworms for decomposing waste remains major concern for the scientists. In order to have quick decomposition of waste materials, DIHAR has introduced *Eisenia foetida* earth worms to the Ladakh region. During winter month the worms kept in Polyench greenhouse to safeguard them from cold injury. The worms kept in greenhouses also enhanced the temperature in the greenhouses through their decomposing activity.

Table 5 : Nutrient Profile of Vermicompost and Farm Yard Manure

Nutrients	Vermicompost	Farm Yard Manure
N (%)	1.6	0.5
P_2O_5 (%)	1.7	0.2
K_2O (%)	0.8	0.5
Ca (%)	0.5	0.9
Mg (%)	0.2	0.2
Fe (ppm)	175.0	146.5
Mn (ppm)	96.5	69.0
Zn (ppm)	24.5	14.5
Cu (ppm)	5.0	2.8
C:N ratio	15.5	31.3

7.1.1 Advantages of Vermicomposting

- It is an ecofriendly natural fertilizer prepared from biodegradable organic wastes and is free from chemical inputs.
- It does not have any adverse effect on soil, plant and environment.
- It improves soil aeration, texture thereby reducing soil compaction.
- It promotes better growth and nutrient absorption.

Table 6 : Recommended Vermicompost dose for different vegetable crops grown at Ladakh

Crop	Vermicompost Dose (ton/ha)
Tomato	10.00
Brinjal	12.00
Capsicum	08.00
Potato	06.00
Cole Crops	09.00
Root Crops	06.50
Onion	09.50
Garlic	11.50
Beans	03.50
Okra	05.50
Cucurbits	08.50
Leafy vegetables	03.50

7.2 Mulching

Mulching is the process or practice of covering the soil/ground to make more favourable conditions for plant growth, development and efficient crop production. Low temperature and partial availability of water due to the complex geographical constraint have adverse effect on vegetable production in the region. Being an arid desert, the region also suffers from low humidity which leads to more evapotranspiration and number of irrigation to the crops. Evaporation from the soil surface may account for as much as 50 percent of the total moisture lost from the soil during the growing season for many agricultural crops (Shaw, 1959). In this context, mulching with plant residues and synthetic materials is a well-established technique for increasing the profitability of many horticultural crops (Gimenez *et al.,* 2002). Such effects are mainly contributed to the capacity of mulch to conserve soil moisture (Vavrina and Roka, 2000) and increase early soil temperature (Shaw, 1959). Irreplaceable water supplies are already being used at an alarming rate as a result water table is going down day by day, and the agriculture is increasingly competing with urban and Industrial water requirements. Moreover, excessive use of herbicides for the control of weeds had led to the problem of herbicide resistance in many crops. In this context, mulching may prove beneficial for crop growth because of complex change in soil environment through modifying soil temperature, reduction in evaporation, weed competition, soil compaction and erosion. The study conducted at research fields of DIHAR revealed that mulching with black polythene of 200 micron thickness conserve moisture and reduces irrigation water requirement up to 30 percent and increases the soil temperature upto 6.5°C depending upon the crop. It is beneficial for early and healthy emergence of the crop during early stage. It create physical barrier for the emergence of the weeds. Mulching stimulates soil microorganisms which results into rapid breakdown of soil organic matter and release of plant nutrients for better crop growth and yield (Table 7).

Table 7 : Effect of polyethylene mulch on yield and earliness of vegetable crops

S No.	Vegetable Crops	Yield (q/ha)		Earliness (Days)	
		Mulched	Non-mulched	Mulched	Non-mulched
1.	Cabbage	780	691	20	26
2.	Cauliflower	516	389	18	24
3.	Capsicum	240	140	22	29
4.	Egg plant	121	58	25	30
5.	Chilli	57	22	25	29
6.	Tomato	617	428	16	22
7.	Bottle gourd	385	235	12	15
8.	Summer quash	560	415	18	23
9.	Okra	128	67	17	22

8. Protected cultivation

The harsh environmental condition of Ladakh does not allow much scope to diversify and increase yield potential of vegetable crops. Instead of taking these as a constraint, DIHAR took solar energy, the most ancient source of energy, as an opportunity to improve the vegetable scenario in Ladakh. DIHAR has standardized various components of vegetable production under protected conditions and developed solar based appropriate and sustainable greenhouse technology for cold desert areas.

(i) *Trench* : It is a very low cost greenhouse termed as an underground greenhouse preferably 30′ x 10′x3′ of size. Trench is covered with transparent UV stabilized 200 micron thick polythene during day time to harvest maximum solar energy and extra cover of black polythene at night time to check the heat loss. It maintains average 7–8°C difference in temperature during winter. The production cost under trench is ₹ 5.25/kg and cost benefit ratio is 1:3.5.

(ii) *Polyench* : It is semi underground double walled greenhouse working on the principle of zero energy chamber made from locally available unbaked bricks and wooden ballies. The air gap between two walls provides better insulation to check the heat loss. It enhances 11–12°C temperature and 40 percent humidity as compared to ambient day time during extreme winters and provide favorable environment for vegetable cultivation especially leafy vegetables and root crops of European origin. The production cost under polyench is ₹ 6.00/kg and cost benefit ratio is 1:4. This technology has been widely accepted by local farmers, and for further impetus, Ladakh Autonomous Hill development Council (LAHDC) Leh is providing 50 percent subsidy to local farmers for construction of such greenhouses.

Several other types of greenhouses such as fiber reinforce polyester (FRP) and polycarbonate (double & triple layered) have also been designed and standardized for high productivity at different altitudes of Ladakh. So far 75 FRP green houses and 120 low cost greenhouses were established by DIHAR along the line of control in army units for round the year availability of vegetables to the troops.

9. Crop Rotation

Crop rotation is the practice of growing a series of dissimilar type of crops in the same space in sequential seasons. The choice and sequence of rotation crops depends on the nature of the soil, the climate, and rainfall, which together determine the type of plants that may be cultivated. A traditional component of crop rotation is the replenishment of nitrogen through the use of legumes in sequence with cereals and other crops.

9.1 Advantages of Crop rotation

- Crop rotation helps to create a biodiverse soil which reduces the chance of soil bound organisms to get a pest and so reduces the use of pesticides.
- It helps to minimise the growth of herbs getting a weed that needs to be eradicated and so reducing the use of pesticides.
- It helps in getting a good soil structure (alternatively shallow and deep rooting roots), high organic matter, good water provision, especially in combination with conservation tillage, resulting in a higher yield.
- Crop rotation, especially performed with nitrogen-fixing rotation crops, will reduce the input of fertilizers and so reduce the pollution by nitrogen.
- Crop rotation, especially combined with conservation tillage, will lead to higher soil-carbon content and so contribute to combating climate change.

9.2 Crop Rotation for Cold Arid Zone of Ladakh

Defence institute of High Altitude Research (DIHAR) has standardised different crop rotation pattern for various solar greenhouses over a period of time which are illustrated below :

Mangol - Tomato - Mangol

Turnip - Cabbage/Cauliflower

Turnip/Spinach - Tomato

Turnip - Cauliflower/Knolkhol - Mangol

Mangol - Onion - Cabbage/Knolkhol

Cauliflower - Carrot

Cabbage - Carrot

Mangol - Brinjal/Capsicum

Radish - Knol-khol/Cabbage

10. Set Technology for Enhancing Onion Yield

Onion is an important crop among the crop grown and supplied to the army in this region. The Local White variety of onion grown by the farmers has low yield and character of splitting. After splitting the bulbs are smaller than the specifications fixed by the army. To find out the solution for this problem large number of onion varieties were evaluated for their suitability at Ladakh region. Among them, Brown Spanish is most suitable because of higher yield and low splitting rate. To enhance yield and availability period of onion in Ladakh region, Sets technology has been introduced and standardized for cold arid climatic condition. Sets are onion bulbs of 15 mm size used for transplanting and production of onion in a particular season.

Fig. 1 : Onion Sets

10.1 Novelty of the Sets Technology

- Sets are less sensitive to adverse soil and climatic conditions then other method of cropping
- High survival rate of the plants
- Suitable for early planting
- More food stuff for vigorous growth
- Less water requirement
- Less damage by the insects
- Suitable for organic production

Higher yield with enhanced availability period can be achieved by cultivating the crop through onion sets.

Table 8 : Comparative survival rate and superiority over traditional method (%)

S. No.	Variety	Survival (%) Traditional	Sets	Superiority over Traditional (%)
1.	Eg-50	52.6	95.4	81.4
2.	Red pinoy	43.3	88.6	104.6
3.	Afdr	38.6	85.7	122.0
4.	Brown spanish	56.5	98.7	74.7

Table 9 : Superiority of yield of onion grown through sets technology on onion grown traditionally grown local variety

Variety	Yield (q/ha) Seedlings	Sets	Superiority (%)
1. Eg-50	616	1268	105.8
2. Local	232	1036	446.6

11. Plant Protection

The long prevailing severe winter of the region is not conducive to survive/spread of many pathogens and insect-pests. But some insect-pests and diseases have been encountered in cultivation of vegetable crops. Health management practices of vegetable crops have been standardized after identification of the major insects pests like cutworm, aphids, cabbage butterfly caterpillar and diseases *i.e.* damping off, black rot, fusarium wilt, powdery mildew, purple blotch and common scab. The intensity of the diseases and insect-pests decreases with the increase in altitude therefore more than 3600 m altitude recommended for healthy seed production of certain vegetables (Garg *et al.,* 2003).

12. Post Harvest Management

The availability of the vegetables for longer period of a year can be maintained by giving interval in sowing/planting, harvesting and storage. Almost all the vegetables are sown/transplanted in the month of May and harvested up to October from the field. They have to be stored to maintain further availability. The vegetables are stored in cold stores at about 4°C temperatures. But unlike plains vegetable storage facilities (cold storage or control environment storage) are not available in cold desert regions. Therefore, zero-energy based underground storage technique has been standardized by DIHAR.

Certain vegetables like, carrot, radish, turnip, cabbage, knolkhol, onion, pumpkin etc can be stored even during frozen winter (November-April) with this technique. The Institute has also developed technology for storage of locally produced potato up to 10 months (November-August) in geothermal energy based underground store with CIPC treatment.

13. Quality assessment

After harvesting, the urgent need is the phytochemical and nutritional investigation of all the agricultural produce so as to have their nutrient profiles. DIHAR with its well equipped laboratory carried out the phytochemical estimation to assess nutritional quality of vegetables (Table 7).

Table. 10 : Nutrient profiling of Radish

Sample/ Parameters	Local Radish (Fresh)	Local Radish (Stored)	Pusa Himani
Fat (%)	0.04	0.15	0.11
Protein (N x 6.25), %	1.10	2.58	0.84
Energy, kcal/100g	28.20	43.19	26.27
Dietary Fiber, %	5.05	8.89	5.32
Calcium, mg/kg	486.83	546.86	248.44
Potassium, mg/kg	2117.98	3245.74	894.1
Sodium, mg/kg	77.47	97.15	37.97
Zinc, mg/kg	2.25	3.94	1.57

14. Vegetable Research and Socio-Economic Upliftment of Local Farmers

The significant findings of the research on vegetables were demonstrated and disseminated to the farmers through various extension programmes. With the adoption of these technologies the farmers are able to produce enhanced quantity of vegetables every year. With R & D inputs and technical guidelines from DIHAR farmers through their cooperative marketing society are able to supply 51 percent vegetable requirement of armed forces locally in 2010. Local farmers are getting a substantial amount (Aproximately ₹ 8 Crore) per annum by selling the vegetables to army other than the civil market and for household consumption.

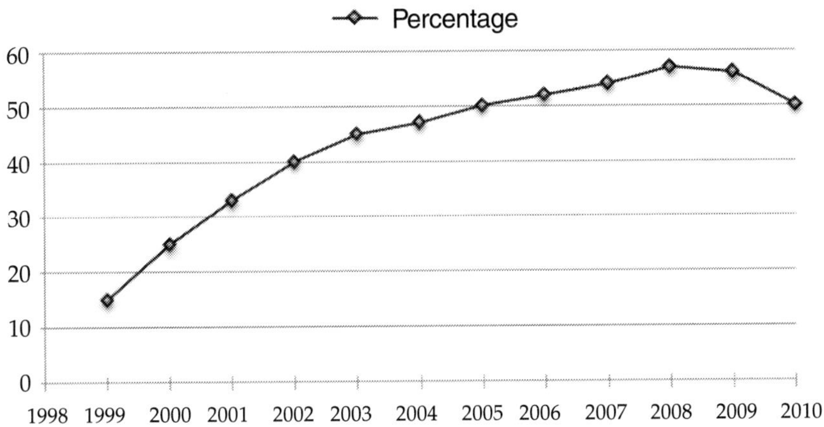

Fig. 2 : Supply of fresh vegetables (% of total requirement) to Army

Conclusion

The research and development activities carried out during last five decades by DIHAR on different parameters of the vegetable crops research provides scientific inputs to the farmers as well as army deployed in the 'L' sector to meet the nutritional requisites and uplift the socio-economic status of the people in high altitude cold desert regions of Ladakh.

References

Bhoyar, M., Mishra, G.P., Singh, R. and Singh, S.B. (2011). Ethno-botany of traditional wild edible plants from cold arid desert of Ladakh-potential source of winter vegetables. *(Indian Forester, A/S (GL) 14/170).*

Bhoyar, M.S., Singh, R., Mishra, G.P. and Stobdan, T. (2009). Lesser-known plants to enrich vegetable basket in Ladakh. *Indian Horticulture,* 54 (4): 16-17.

Chaurasia, O.P., Singh, N. and Singh, B. (2000). Potential wild leafy vegetable from Ladakh. *Indian Horticulture* (July- September). 4-5.

Garg. I.D., Chandla, V.K., Paul Khurana, A.M. and Singh, N. (2003). Potato cultivation and health status of the crop in Leh-Ladakh (J&K). J. Indian potato Assoc. 30(1-2): 121-122.

Gimenez, C., Otto, R.F. and Castilla, N (2002). Productivity of leaf and root vegetable crops under direct cover. *Scientia Horiculturet,* 94: 1-11.

Shaw, R.H. (1959). Water use from plastic covered and uncovered corn plots. *Agronomy Journal,* 51: 172-173.

Singh, N., Rout, B. and Kumar, N. and Mohanty, J.N. (2003). Off season potato production in trench in cold arid conditions of Ladakh. Journal of Indian Potato Association. 30 (1-2):95-92.

Singh, N., Rout, B. and Kumar, N. (2003). Cultural tips for growing vegetable in Ladakh, Field Research Laboratory, Leh-Ladakh, India.

Singh, N. and Ahmed, Z. (2005). Solar greenhouse technology for vegetable production in high altitude cold arid region. Field Research Laboratory, Leh-Ladakh, India.

Vavrina, C.S. and Roka F.M. (2000). Comparison of plastic mulch and bareground production and economics for short-day onion in a semitropical environment. *Horticultural Technology*, 10: 326-330.

□□□

Chapter-2

Solar Greenhouses : A Boon for Enhancing Vegetable Production in Sub Zero Climatic Conditions

Pal Murugan M.[1], Narendra Singh[1], Dorjey Angchok[1], Manish Bhoyar[1], A.K. Katiyar[1], R.P. Bhatt[2], Hitesh Kumar[1] and R.B. Srivastava[1]

[1]Defence Institute of High Altitude Research (DIHAR), DRDO, Leh, Ladakh

[2]Directorate of Life Sciences, HQ DRDO, Delhi

1. Introduction

Food and nutritional security remain the major constraints in the Trans Himalayan region which stretches from Tajikistan in the West to Bhutan in the East, with the Tibetan Plateau in the centre and circumpolar region of the world due to high altitude and arctic-like condition. Ladakh (Jammu & Kashmir), Lahul & Spiti (Himachal Pradesh) and some parts of Uttarkhand are the high-altitude cold desert regions of India which fall under Trans Himalayan zone. Climate of these regions is very harsh: the prevalence of extreme sub-zero cold temperature between October to April (-10 to -40°C) and scorching heat during July to August low precipitation (20-30 cm) in the form of snow, low relative humidity (20-45%) and high wind velocity (70 nautical mile/hr). These conditions result in very sparse vegetation and arctic like desert conditions, which makes the cold desert an inhospitable terrain for crop cultivation (Pal Murugan *et al.*, 2010). As natural resources are limited, the local population has to primarily rely on sheep, goats, cattle and yak as their main source of survival, with limited substainable agriculture on very small landholdings focused mainly on cereals.

Besides the local population, a large number of Army and Para military personnel are deployed in this region due its strategic location as it shares border with Pakistan and China. This adds to the huge amount of pressure on the food and nutritional security of this region. Besides Army and the floating population, vegetables are required for a population of 3.0 lakh in the region (Singh *et al.*, 2008). If it is calculated that total of 37,200 metric ton/year vegetables are required (Table 1). Out of this quantity, approximately 17,705 metric ton/year vegetables are produced locally and remaining quantity is transported from other places of the country.

The sub-zero temperature prevalent in Ladakh during the winter months (Oct-April) restricts the crop cultivation to a single season which is mainly focused on wheat and barley with some turnip, radish, cabbage potatoes, peas and occasionally other vegetables between May to September. The valley is devoid of cultivation during rest of the year, which necessitates storage of excess produce from the cultivated field for consumption in the winter. The local people and Armed forces in Ladakh have to depend upon nearby region i.e. Chandigarh and other parts of Jammu & Kashmir for the supply of fresh vegetables and other food items to meet their daily requirements.

Table 1 : Total requirement and local production of vegetables in Ladakh

Items name	Requirement (MT/year)	Production (MT/year)
Vegetables	17189	10378 (>30 types)
Fruits	9925	2741
Potato	16120	8259
Onion	7759	3098
Garlic	1388	275
Total	52381	15410

To boost the production of plant-based food in the region, DIHAR has developed several agro-technologies to grow more than 78 types of vegetables in this region, such as suitable hybrids and varieties, their cultural techniques, mulching, storage, value addition and post harvest management, organic farming, vegetable seed production and solar greenhouse technologies.

2. Greenhouse Cultivation

Greenhouse cultivation or Protected cultivation practices can be defined as cropping techniques wherein the micro climate surrounding the plant body is controlled partially/fully, as per the requirement of the plant species grown, during their period of growth. It is used to protect the crops from adverse climatic conditions and to provide optimum conditions of soil, light, temperature, humidity carbon dioxide etc. for the best growth of the plants to achieve maximum yield coupled with quality. A greenhouse is a framed or inflated structure covered by transparent cladding material which protects the plants from high wind velocity, excess evaporation, unusual precipitation, and temperature extremes. Depending upon the cladding material and its transparency, major fraction of sunlight is absorbed by vegetable crops and other objects. These objects in greenhouse in turn emit long wave thermal radiations for which cladding material has lower transparency. With the result, solar energy is trapped and it raises the temperature inside the greenhouse. This is popularly known as greenhouse effect. The CO_2 released by plants during nighttime is also trapped inside the greenhouse which is subsequently taken up by the plants itself during daytime in the presence of light, and thus enhances photosynthesis. The above factors contribute to 3-5 times higher yield of crops under greenhouse condition, as compared to open-field conditions. With the advancement in agriculture, various types of protective cultivation practices suitable for a specific type of agro-climatic zone have emerged. All greenhouses collect solar energy. Based on the way of using solar energy, greenhouses are grouped into two types:

2.1. The Active Solar Greenhouses

The active solar greenhouses comprise of a collector (mirrors and mirror plates) system to capture solar radiation. The collectors convert the harnessed solar radiation into heat. The collector is designed in the form of a box or a pipe, which contains water, anti-freeze substances or air. As the solar radiation falls on the mirror plates, it is directed to the collector, and solar energy is produced. The air or the liquid present in the collector gets heats up because of the solar energy. The heated gas or liquid from the collector is transferred to various parts of the greenhouses by using a pump or a venting mechanism, and this helps in heating up the entire structure. With use of electricity from outside sources the required climatic conditions for crop can be created in greenhouse.

2.2. Passive Solar Greenhouses

Passive solar Greenhouses use the architectural design and construction material of a building to produce solar energy. The building, along with its thick, absorbent walls, functions as the collector of solar radiation during daytime. The building material absorbs heat, retains it throughout the day, and starts emitting it during evenings, providing heat to the building when the sun sets. A greenhouse is another classic example of passive solar heating systems.

Greenhouse effect is being used since long in many places to grow various crops under controlled environmental condition by using electricity. However, in high altitude cold arid regions production of hydro power is not easy especially during winter. Due to subzero temperature most of the water sources freeze and the use of fossil fuel generated electricity is also not economical and environmental friendly (Singh and Ahmed, 2007). Therefore, development of suitable passive solar-based greenhouse is necessary as abundant sunshine and high intensity solar radiation is available in the high altitude regions. Passive solar greenhouses are often good choices for small growers, because they provide a cost-efficient alternative for farmers to extend the growing season.

3. Concept and Principle of Passive Solar Greenhouses

Passive Solar greenhouses or solar greenhouses in short are designed to collect solar energy during sunny days and store heat for use at night or during periods when it is cloudy. A solar greenhouse may be an underground pit, a shed-type structure, or a Quonset hut. Large-scale producers use free-standing solar greenhouses, while attached structures are primarily used by home-scale growers (Alward *et al.*, 1981).

The successful functioning of the solar greenhouses depend upon following factors:

1. Effective collection of the maximum amount of solar radiation during the day.
2. Efficient storage of the heat collected from the sun radiation during the day.
3. Insulation in the greenhouses to avoid heat loss.
4. Release of this heat to the inside of greenhouse during the night.

3.1. Collection and Storage of Radiation

3.1.1. Solar orientation

Solar orientation plays a major role in heating the greenhouses. Since the sun's energy is strongest on the southern side in the Trans Himalayas region, glazing for solar greenhouses should ideally face true south. However, if trees, mountains, or other buildings block the path of the sun when the greenhouse is in a true south orientation, an orientation within 15°C to 20°C of true south will provide about 90% of the solar capture of a true south orientation (Alward *et al.*, 1981). The latitude of location and the presence of potential obstructions may also require adjusting the orientation of greenhouse slightly from true south to obtain optimal solar energy gain.

3.1.2 Glazing materials

In order to get maximum sunlight the glazing material of greenhouses should be south faced. Glazing materials used in solar greenhouses should allow the greatest amount of solar energy to enter into the greenhouse, while minimizing energy loss. In addition, good plant growth requires that glazing materials allow a natural spectrum of photo-synthetically active radiation (PAR) to enter. Rough-surface glass, double-layer rigid plastic, and fiberglass diffuse light, while clear glass transmits direct light. Although plants grow well with both direct and diffuse light, direct light through glazing subdivided by structural supports causes more shadows and uneven plant growth. Diffuse light passing through glazing materials evens out the shadows caused by structural supports, resulting in more even plant growth (Giacomelli and Gene, 1999 and Bartok *et al.*, 2001).

3.2. Storage, Release and Containment of Heat

Different materials are used for the storage and release of heat in the solar greenhouses according to the availability and greenhouse design. The most common method for storing solar energy is to place rocks, concrete, or water in direct line with the sunlight to absorb its heat. In Ladakh temperature during night hours may drop down to below zero degree Celsius if heat observed from solar radiation is not properly conserved. The lists of material are summarized below to maintain thermal environment of greenhouses.

3.2.1. Opaque materials

These materials block solar radiation but they can allow transfer of energy by heat conduction. There are two main types.

3.2.1.1. Dense materials

Dense materical such as mud bricks, black stone, and cement and clay soil can conduct and store heat. Except for metals, heavier materials generally store more energy and absorb it faster. In a greenhouse, dense materials are used to provide thermal mass and as load-bearing materials to form the wall supporting the polythene frame and roof. Mud Bricks are the mostly used material for construction in Ladakh condition. For proper insulation, the walls made of mud bricks should be made in east, west, and north sides. A solid roof should be provided on the north side to limit heat loss. The roof is tilted to avoid shading in winter, reduce the interior volume and easy removal of snow fall. In case of low cost greenhouse such as Polyench and Ladakhi poly house the structure should be double walled. The walls are composed of three layers: an outer load-bearing wall built with mud brick, plastered by clay soil, or stone; an inner wall used to store heat during the day and release it at night, also built with mud brick, rammed earth, or stone; and an insulating layer of materials like straw, sawdust, wood shavings, dry leaves, dry grass, or wild bush cuttings pressed between the two. In case of high cost greenhouse such as fiber reinforced polyester greenhouse and polycarbonate greenhouse the wall height will be 2 feet from the ground level and no need to have double wall structure as the sheets itselves have the properties of both heat conduction and storage. Black colored stones can be placed in greenhouses to maintain the heat.

3.2.1.2. Low density materials

Light materials like straw, sawdust, wood shavings, wooden poles and dry leaves, and dry grass, are poor conductors and storers of heat and are thus good insulators: they help retain the heat inside the greenhouse. These materials are filled into the cavity between the load bearings wall and the thermal mass wall in low cost greenhouses such as Polyench and Ladakhi Polyench for energy saving.

3.2.2. Wall colour

The color of the thermal wall also plays a major role in slow release of heat to the greenhouse during the night. The inner side of the west wall is painted white to reflect the morning solar radiation after the coldness of the night; the inner side of the east and northern wall is painted black to absorb and store the afternoon solar radiation, which is then released at night to heat the interior space.

3.2.3. Door and ventilation

The door is located on the wall opposite to the side from which the prevailing wind blows (the lee side) to reduce infiltration of cold air. On bright sunny days, the air in the greenhouse can become very warm. Overheating (over 30°C) can damage the vegetables and encourage diseases and pests. Manually operated openings (ventilators) are provided in both sides of (door, wall shutter). The warm air rises and leaves the greenhouse through the roof ventilator, drawing in the cooler ambient outer air through the lower ventilators.

4. Infra-structure of Greenhouses

4.1. Glazing Material

For the efficient collection of solar radiation transparent solar absorbing materials like single and factory sealed double glass, transparent single and double layered polyethylene polythene sheet, acrylic sheets, fiber reinforced polyester sheet and polycarbonate sheets can be used to heat the greenhouses depending upon the economic condition of farmers. The method used for mounting the glazing material affects the amount of heat loss (Bartok and John, 2001). For example, cracks or holes caused by the mounting will allow heat to escape, while differences in the width of the air space between the two glazes will affect heat retention. Installation and framing for some glazing materials, such as acrylics, need to account for their expansion and contraction with hot and cold weather (Giacomelli and Gene, 1999). As a general rule, a solar greenhouse should have approximately 0.75 to 1.5 square feet of glazing for each square foot of floor space. This warms the interior space and is absorbed by the vegetables, the ground, and the walls.

Plastics are the dominant type of glazing materials used in greenhouses, with the weatherability of these materials being enhanced by ultraviolet radiation degradation inhibitors, infrared radiation (IR) absorbency, anti-condensation drip surfaces, and unique radiation transmission properties (Giacomelli and Gene, 1999). The glazing materials can be covered with a moveable layer of insulation like a curtain, cloth, or mat opaque black polythene sheets after sunset to reduce night time heat loss. The Glazing materials should be mounted on roof of greenhouse at an angle of 35-45 degree to absorb maximum.

4.2. Construction Materials

Besides glazing material, several other materials are also required for construction of greenhouses. It is better to prefer the use of locally available material for constructions which not only reduces the installation cost but also helps to popularize the technology. The major construction material required for greenhouses is its frame structure which could be either of metal or wooden. Iron angle. GI pipes or aluminum frames are the most common metals used in greenhouses. Wooden frames can also be used which are locally available in cold arid region. Wooden frames should be prepared in cold arid region since metals lost heat very quickly as they are the good conductors of heat. Strong foundation of frame is essential in cold arid region to resist the strong winds. Stability is further ensured by constructing earthen brick wall at the back and on either sides (Brahma Singh *et al.*, 2000).

5. Principles of Site Selection

- The site should be near to the water source as vegetables and fruits require frequent irrigation to grow.
- The site should be free from any shade, trees and obstacles as they block the sunlight.
- The land should be flat and dry and south faced. If it is sloppy, north faced greenhouses should be avoided.
- It should be fenced as stray cattle menace is the biggest problem during winter in Ladakh.
- In heavy snow fall area, the roof should be steep angled to avoid snow load.

Keeping the above principles, climatic conditions and resource condition of farmers following free standing solar greenhouses *viz,* Polyhouse, Trench, Polyench, Polytunnels, Fiber Reinforced Polyester (FRP) and Polycarbonate Greenhouses suitable for Trans-Himalayan region were designed and developed by DIHAR over a period of five decades. These greenhouses are entirely heated by sunlight and effective for cultivation of vegetables and fruits throughout the year with no additional fuel based heating as the availability of electricity is the biggest constraint in the Trans-Himalayan region. These greenhouses are adopted by small and marginal farmers and armed forces to meet their nutritional requirement.

6. Types of Greenhouses in Ladakh

6.1. Glasshouse

Glass panels with light transmission percentage of 70-90% (single or factory sealed double layered) are used as a glazing material in the greenhouse. The glass houses are constructed in all shapes and sizes and are quite effective for winter cultivation. The variation in temperature between outside and inside conditions is 20 to 25°C. The ideal size for Trans-Himalayan region is 30 feet length and 10 feet width with centre height of 7 feet and side height is 6 feet. This greenhouse is also useful for early crop production and early nursery raising. High initial cost, difficulty in construction and frequent damage of glass panels and requirement of heavy support are limiting factors in development of glass houses in Ladakh regions.

6.2. Polyhouse

Ideal features of polyethylene have increased the use of polyhouses in place of glass houses throughout the globe. It has not only reduced the initial cost but also increased the popularity of greenhouses by simplifying the installation technology. Generally two types of polyhouses are there in Ladakh.

Fig. 1 : Glasshouse

Fig. 1 : Polyhouses

6.2.1. Prefabricated polyhouses

A prefabricated metal structure of convenient size is installed and protective shade made up of polyethylene sheets either double or single layered with light transmission percentage from 60-90 percentages is used for growing high-value agricultural crops. The structure is prepared with the PVC or iron pipes. The polyhouses can also be constructed with the help of ultraviolet stabilized plastic sheets, so that they may last for more than 5 years. It is semi-circular, square or elongated in shape. The ideal size for Trans-Himalayas is 30 feet Length and 10 feet width with centre height of 7 feet. Heat loss can be significantly reduced using a blower. Treatment with ethyl vinyl acetate enhances resistance to cracking in the cold and tearing. Crops grown in the polyhouses are protected from intense heat, bright sunlight, strong winds, hailstones and cold waves. Poly houses are used for early nursery rising for summer production and growing high value vegetables during summer. The major disadvantages of poly houses are that they are very easily torn, light transmission decreases time and their very low life span. However, due to the prevalence of subzero temperature, success in the poly house crop production is very low due to poor temperature retention and high installation cost (Singh *et al.*, 2008).

6.2.2. Ladakhi Polyhouse

This is the low cost greenhouse most commonly found in Ladakh and is made up of locally available materials. It is similar to normal polyhouses, but the only difference is its surrounding mud bricks wall in place of polythene sheets which not only cut down the installation cost but also safegaurd the greenhouses from adverse effects of winds.

The wall increases temperature retention. The ideal size of this greenhouse is 30 ft x 10 ft with center height of 7 ft. Polyethylene sheet is the major glazing material used for heating this greenhouse. The height of back wall should be 7 ft with side walls slanting towards the front wall which is at one feet height. The low front walls are placed inside the greenhouse near the south-facing windows. They absorb heat on the front (south-facing) side of the wall and then radiate this heat into the greenhouse through the back (north-facing) side of the wall, which facilitates more insulation inside the greenhouses. The wall facing the northern side can be single or double walled and painted with black color. If it is double walled, straw or saw dust or wood shaving and waste leafy materials can be filled within the four inch gap between two walls in order to reduce the heat loss. This greenhouse has boosted the greenhouse revolution in Ladakh.

Fig. 3 : Ladakhi Polyhouse

6.3. Low Poly/Fiber Glass Tunnels

These are semispherical structure frames made of metal, wooden logs, other locally available materials and PVC pipes and covered with a transparent material such as polythene and fiber reinforced plastics with size of 6 feet length x 3 feet width with center height of 2.5 feet to create protected environment. These structures serve as mini-greenhouses. This is a temporary structure which can easily be shifted to any place. It is generally used for raising early nurseries and winter vegetable production. It can also be used in open field during the early summer and late summer to protect the crop from cold and chilling injuries. However, major problems in tunnels are frequent removal and replacement of polythene sheets (cover) for intercultural operation such as irrigation weeding etc.

Fig. 4 : Poly tunnels

6.4 Trench Greenhouse

It is a very low cost greenhouse termed as an underground greenhouse. This is the simple, cheap and useful greenhouse structure for Ladakh. It is also called as resource poor farmers greenhouses. The preferable size is 30′ x 10′x3′. Trench is a pit covered with transparent UV stabilized 200 micron thick polyethylene sheets during day time to harvest maximum solar energy and extra cover of black polythene at night time to check the heat loss. The damage of blowing of polyethylene film by strong wind is minimized by putting the stones along the sides. This structure does not require much skill in its construction and management. It maintains average 7-8°C difference in temperature during winter. It uses both solar and geo thermal energy for growing vegetables. The walls inside the pit are made up of stones and plastered with either cement or clay soil. The trench greenhouse should be made on north-south direction to get maximum sunlight. It is the cheapest greenhouse that can be made with family members and caters the vegetable needs of small family.

Fig. 5 : Trench Greenhouse

6.5. Polyench

It is a semi-underground double walled greenhouse working on the principle of zero energy chamber made up from locally available unbaked bricks and wooden poles and a hybrid structure between and trench and poly house greenhouses. The greenhouse is basically a trench of 10m length, 3m width and 1.0m deep. The orientation is such that longitudinally it runs east to west. The width of foundation walls is 1.0m and depth 30cm below level of trench. Stone masonry walls are constructed on all sides from excavated level to the ground level over foundation. The width of stone masonry wall is 0.75m and height is 1.0m above foundation. Over this wall, above ground level (GL) mud brick masonry double wall of total thickness 0.75m including gap of 15cm in between, are constructed. The height of mud brick wall above GL is 2m in north side and zero meters in opposite side (Southern side). Similarly, side walls are also constructed in that way and the roof is provided with mentioned degree of slope. The roof consists of 2 parts, of which 1 part is covered with polythene sheet and the other with mud covering. The polythene roof is provided with slope of 2.0m height in 2.5m length towards southern walls (short wall). Mud roof is provided with slope of 0.5m height in 1.5m length towards Northern side wall (higher wall). Wooden door of size 2m height and 1m width is provided on the Eastern side of the wall under mud roof covering 03 numbers of ventilators of size 2.0ft by 1.5ft are provided at roof near highest level. Two ventilators of same size (2.0ft x1.5ft) are provided on the Western side wall.

An irrigation inlet is also provided in eastern side wall. The air gap between two walls provide is better insulation to check the heat loss. It enhances 11-12°C temperature and 40% humidity, as compared to ambient day time, during extreme winters and provides favorable environment for vegetable cultivation especially leafy vegetables and root crops of European origin. The production cost under Polyench is 6.00 Rs/kg and cost benefit ratio is 1:4. This technology has been widely accepted by local farmers, and for further impetus Ladakh Autonomous Hill development Council (LAHDC) Leh is providing 50% subsidy to local farmers for construction of low cost greenhouses (Singh and Ahmed, 2007).

Fig. 6 : Polyench Greenhouse

6.6. FRP Greenhouse

The glazing material used in the greenhouse is fiber reinforced polyester sheets. The normal dimensions of the greenhouse are (100 ft x 30 feet) with a centre height of 10 feet and a side height of 6 feet. The translucent nature of this material diffuses and distributes light evenly with light transmission percentage of 85-90 and can lost up to 15 years. The initial cost required for fabrication of the greenhouse is high, but the comparative life of the greenhouse is much more than others. It is constructed on the structure made of iron/aluminum structure and designed domed-shaped or cone-shaped. Additional insulation can be provided by establishing trenches and poly house in the FRP greenhouse. This greenhouse can be effectively used for high value crops production of indeterminate hybrid of tomato, capsicum, chilli, brinjal, gynacious hybrid of cucumber and zucchini, during summer months from 2,400-3,600 m altitude and leafy vegetables, i.e., swiss chard, beet leaf, kale, lettuce, etc. during winter. Light transmission decreases over time; poor weather resistance, most flammable of the rigid glazing materials and insulation ability which do not cause snow to melt, are some of the disadvantages of FRP Greenhouse.

Fig. 7 : FRP Greenhouse

6.7. Double and Triple Layered Polycarbonate Greenhouse

The glazing material used in the greenhouse is double layer polycarbonate sheets. Polycarbonate sheets are famous for its qualities such as resistant to fire, resistant to UV radiation, strong and light weight, easy to cut and install. The life span of sheet varies from 15-20 years. Normal dimensions of the greenhouse are (120 ft x 30 feet) with a centre height of 11 feet m and a side height of 6 feet. The variation in temperature between outside and inside conditions is 10-20°C. The life span of polycarbonate sheets is upto 20 years. However, cost of this greenhouse is much higher and not suitable for small and medium farmers. It can be effectively used for extreme altitudes also. Triple layered polycarbonate greenhouses are most suitable to grow early seedlings i.e. in 1st week of April for protected cultivation to take early crop and also ensuring 300 per cent cropping intensity.

Fig. 8 : Polycarbonate Greenhouse

7. Solar based thermal greenhouse

As the passive solar greenhouses designed and developed by DIHAR are not able to maintain the temperature more than 15°C during winter months where outside temperature dips up to -30°C, the feasibility of using active solar greenhouses was tested at Defence Institute of High Altitude Research 2010-11. This greenhouse is entirely heated by solar energy through air and water / Thermic fluid heater during day and night, respectively. The evacuated tube collector with thermal fluid and air is used for collection of solar energy. The size of greenhouses is 45 feet X 15 feet X 9 feet with active radiation of 1200-1400 micromoles/sec./m^2 through CFL/LED bulbs. The temperature, light and humidity are controlled through sensors. The technology is under evaluation at the institute.

Fig. 9 : Solar thermal greenhouses

8. Cost of Production in Greenhouses

Economics is also an important factor to utilize the greenhouse for different aspects of vegetable production. DIHAR has evaluated and recommended economically viable crop rotations and production techniques of the crops for different greenhouses. Tomato (indeterminate hybrid) during summer months (April-October) and Swiss chard during winter month (November-March) give highest return (1:1.8 benefit ratio) in FRP greenhouse, however, leafy vegetables like beet leaf, swiss chard and kale during winter months (September-May) and radish during summer months (June-August) found most viable crop rotation for Polyench (1:3.2 benefit ratio). The data presented in the figure 10 revealed that double and single wall polyenches are found to be most effective due to their low construction cost and higher production potential followed by trench (Singh *et al.*, 2008).

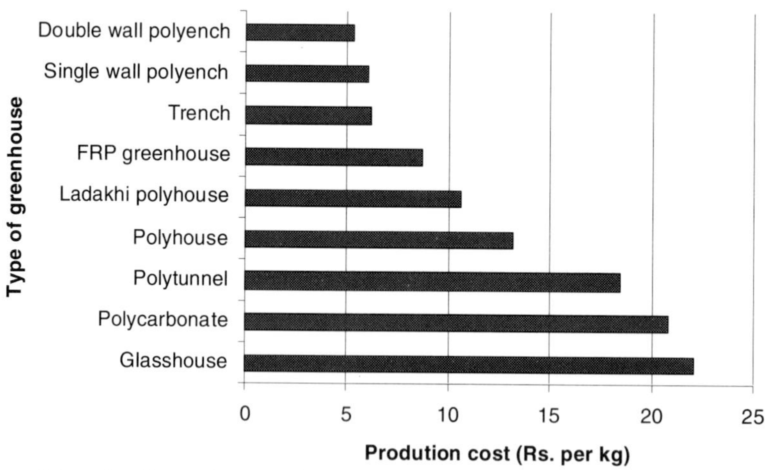

Fig. 10 : Cost of vegetable production in various greenhouses

9. Agro Techniques for Crop cultivation in Greenhouses in Trans Himalayas

Among all the agro techniques, selection of crops according to climate and temperature play a major role in crop production. The passive solar greenhouses developed by DIHAR will not facilitate the growth of high temperature loving crops such as Tomato, Cucurbits and other flowering crops during winter months. The suitability of crop plants according to season have been identified by DIHAR.

9.1 Vegetables Suitable for Winters

Mostly leafy vegetables are suitable to be grown in all types of solar based greenhouses during winter months (Oct-April). But yield of the vegetables depend upon the altitude, type of the greenhouse and selection of the crop to be grown in greenhouse. *Amaranthus*, beet leaf, endive, fenugreek, karamsag, onion, garlic, radish, spinach and swisschord can be effectively grown during winter months by providing proper insulation.

9.2 Vegetable Suitable for Summers

The high value vegetables which are difficult to be grown in open field during summer season can be grown effectively in greenhouse in order to get maximum yield. Vegetable crops such as tomato (determinate and indeterminate type), capsicum, chilly, brinjal cucurbitaceous crops, ladies finger, early (off season) potato will be effectively cultivated in the solar greenhouses.

9.3 Seed Sowing

Timely sowing of seeds plays an important role in obtaining higher yield. Crops like spinach, mustard, fenugreek, chenopodium, amaranthus can be sown directly in pre-prepared beds inside the greenhouses. Cole crops such as cabbage, cauliflower, capsicum, chilly, brinjal cucurbitaceous crops such as summer squash bottle gourd and others are sown in nursery beds and transplanted in the greenhouses. These crops require 25-60 days depending upon the crop and weather condition for raising of the seedlings.

9.4 Soil Preparation

The soil should be finely tilled and well leveled after two to three ploughing of the beds before sowing in order to support an early and even germination and faster growth.

9.5 Manures and Fertilizers

To make the soil fertile enough, well decomposed organic manure at the rate of 3-5 kg per sq meter should be mixed in the beds of the greenhouse. In addition 2 kg per sq meter should be given after 30-50 days of sowing/transplanting of the crop. Yield may be increased with the application of chemical fertilizers e.g Urea, Diammonium Phosphate (DAP) and Minate of Potash (MOP) especially for hybrid crops which require heavy dose of nutrients.

9.6 Plant Population

To get maximum yield, plant population should be maintained inside the greenhouse by sowing/transplanting of the seeds/seedling at proper distance.

9.7 Irrigation

Irrigation should be given to vegetable crops at optimum moisture level. Generally there is not much need of irrigation during winter season in the greenhouses. Once field is well moist then irrigation required at 15-20 days interval.

9.8 Weeding and Hoeing

To keep the beds free from weeds and proper plant growth 2-3 weeding are required during the life cycle of a crop.

9.9 Plant Protection

Incidence of pest and disease is very low due to harsh climatic condition prevailing in Ladakh. Aphids, red spider mite and cabbage caterpillar are the major pest of Ladakh. These insects can be controlled by spraying 2 percent Biolep and 2 percent neem gold the herbal insecticides (Singh and Ahmed, 2005).

10. Greenhouses at LOC

In order to meet out the fresh vegetable requirement of armed forced deployed in border area, DIHAR has established 72 FRP greenhouses, 65 low cost trench greenhouse and 37 Polyench type greenhouses along the line of control in Ladakh region. Besides this on farm demonstration, in-house training and inputs such as seeds and seedling are also provided to have nutrient rich vegetables in the remote location.

11. Benefits of Greenhouse Technology in Ladakh

The benefits which can be derived from the greenhouse cultivation are as follows:

- Environment control allows raising plants anywhere in the world at any time of the year. With the use of greenhouses, cultivation of vegetables such as swiss chard, spinach, coriander, cabbage cauliflower and fenugreek during extreme conditions of winter (October–April) are possible when it is not possible to grown in open field.
- The high-priced vegetables such as cucurbits, capsicum, brinjal, tomato and cucumber which are rare in Ladakh can be effectively cultivated in greenhouses during summer.
- It facilitates the early crop production and useful for early nursery raising. In the Trans–Himalayan region season for open field cultivation starts between mid Aprils to 2 week of May only. Raising seedling in greenhouse in March to April for transplantation will favor early crop maturity.
- It also favors off season vegetable cultivation thereby provides employment opportunities during winter times.
- It is also helpful in maintaining the purity of the seed and useful for conservation of plant species as many plant species in the Trans Himalayas are under endangered condition.
- The control of the microcosm allows the production of higher quality products which are free from insect attack, pathogens and chemical residue.
- High value and high quality crops could be grown for export markets. Income from the small and the marginal land holdings can be increased by producing crops meant for the export markets.
- It can be used to generate self employment for the educated rural youth in the farm sector.

12. Future Prospects

The passive solar greenhouses developed and designed by DIHAR are effective in high value vegetables production between May–Sep and for growing leafy vegetables during the extreme subzero temperature (- 10°C to -30°C). The experiments conducted in polyench and trench greenhouses during 2009 to 2011 by inoculating vegetable crops with *Piriformospora indica,* (a multifunctional symbiotic fungus) showed that vegetables such as cabbage, cauliflower, pea, broad bean carrot, beetroot, knol khol, onion, turnip, and endive could be produced during extreme winter also. So identification of such growth promoting microorganism will favor all type of vegetable production under passive solar greenhouses during subzero temperature. As it is difficult to maintain required environment such as temperature, humidity, light, etc. for growing different vegetable crops in the passive solar greenhouses, development of active solar greenhouses is necessary for socio economic development of Ladakh region. As Ladakh has 290-310 bright sunny days/year (with average solar radiation 195 KWh/m^2/month,average wind speed : 70 nautical mile/hr and hot springs in different region of Ladakh, use of solar, wind and geothermal energy can be effectively used for heating , maintaining the temperature and generation of electricity to have active solar greenhouses. So tapping of nonconventional energy source for creation of active greenhouses will be useful in meeting the food and nutritional security in the Ladakh region. With the help of agro technologies developed at DIHAR including passive solar greenhouses, 50 per cent fresh food requirement of Ladakh inhabitants has been achieved. Active greenhouses with non conventional energy sources coupled with passive solar greenhouses and agro technologies of DIHAR are the way to achieve further more fresh food requirements in Trans-Himalayan region.

References

Alward, Ron, and Andy Shapiro. (1981). Low-Cost Passive Solar Greenhouses. National Center for Appropriate Technology, Butte, MT. 173 p.

Bartok, Jr. and John, W. (2001). Energy Conservation for Commercial Greenhouses. NRAES-3. Cornell University, Ithaca, NY. 84 p.

Brahma Singh, Dwivedi, S.K. and Sharma, J.P. (2000). Greenhouse Technology and Winter vegetable production in Cold Arid Zone.In Dynamics of cold arid Agriculture, edited by Jag Paul Sharma and Aziz Mir A,Kalyani Publishers,New Delhi.

Giacomelli, Gene, A. (1999). Greenhouse glazings: Alternatives under the sun. Department of Bioresource Engineering. Cook College. Rutgers University. Accessed at: http://AESOP. RUTGERS.EDU/~ccea/publications.html

Pal Murugan, M., Janifer Raj, X., Phani Kumar, G., Sunil Gupta and Shashi Bala Singh. (2010). Phyto foods of Nubra Valley, Ladakh -The Cold Desert. *Indian Journal of Traditional Knowledge*, 9 (2): 303-308.

Singh, N. and Ahmed, Z. (2005). Solar greenhouse technology for vegetable production in high altitude cold arid region. Field Research Laboratory, Leh-Ladakh.

Singh, N. and Ahmed, Z. (2007). Double walled Polyench greenhouse. Indian Horticulture, May-June, 2007, 8-9.

Singh, N., Singh, R., Kumar, H., Bhoyar, M. and Singh, S.B. (2008) Sustainable vegetable research for nutritional security and socio-economic upliftment of tribals in Indian cold deserts. In Advances in Agriculture, Environment and Health, edited by S.B. Singh, *et al.* SSPH Publication, New Delhi.

❑❑❑

Chapter-3

Seed Production of Biennial Vegetable Crops in Trans-Himalayas

Narendra Singh, Dorjay Angchok, S.B. Maurya and R.B. Srivastava
Defence Institute of High Altitude Research (DIHAR), DRDO, Leh, Ladakh

1. Introduction

Plant growth, seed yield and quality are greatly influenced not only by genetic factors but also by climatic conditions, such as light, temperature, rainfall, humidity, wind, disease and pests. For seed production of certain vegetable crops, climatic conditions of Ladakh are very suitable. The region enjoys the availability of long photoperiod, high light intensity, low rainfall and humidity, low disease and pest incidence (Garg *et al.*, 2003). However, with above mentioned positive climatic factors, crop specific agro-techniques are essentially required for successful seed production of any crop. Annual vegetable crops like pea, french bean, broad bean, tomato, coriander, summer squash etc. complete their life cycle from seed germination to seed formation in one cropping season. With slight modification in agro-techniques, method of seed production of annual crops is almost similar to as follows in plain conditions because, before the onset of severe winter, these crops complete their life cycle. Biennial vegetable crops like cabbage, knolkhol, onion, beetroot, radish, temperate carrot, turnip, swede, celery, parsley etc. complete their life cycle in two cropping seasons (Singh N. and B. Singh, 2000). In first season they produce vegetative parts and in subsequent year/season, the plants produce flowers and seeds. The seed production of these crops involve several steps such as production of vegetative part in first year, storage of vegetative part during ensuring winter, re-planting of crop in next season in field, roguing and suplementation of specific agro-techniques for successful seed production.

Temperate climatic conditions are essentially required to produce seed of biennial vegetable crops. In India, seeds of these crops are produced only in hilly areas like Kullu, Katrain, Kalpa, Lahaul Spiti, Kashmir valley and Nilgri Hills. Plants of the temperate biennial crops do not bolt until they have met essential chilling requirement. An exposure to low temperature of 4 to 7°C for 6 to 8 weeks initiates bolting in these crops (Singh *et al.*, 1997). Ladakh having temperate climatic conditions can be exploited for production of seeds of these crops. Defence Institute of High Altitude Research (DIHAR) has developed technologies for quality seed production of certain temperate vegetable crops in high altitude cold desert climatic conditions for Ladakh. The biennial vegetable crops such as cabbage, knol-khol, onion, leek, beetroot, temperate radish, temperate carrot, turnip, swede, celery, parsley can be grown in sandy to heavy soils with adequate water and nutrient supply. The yield of bulbils of garlic can be prodiate of through flouring index as deseribed by Mathew, 2005. All the fields where vegetables, wheat and barely are growing in Ladakh, are suitable for seed production of these crops. The techniques developed at DIHAR for seed production of certain biennial vegetable crops are described below.

2. Seed Production of Cabbage and Knol-khol

There are three distinct methods of seed production of these crops viz head/knob to seed method, seed to seed method and the stump method, in agro-climatic conditions of Ladakh, head to seed method for cabbage and knob to seeds for knolkhol are the most suitable. Seed produced from these methods will be of good quality because the heads/knobs provide a better scope for inspection, selection and roguing.

2.1. First Growing Season

2.1.1 Raising of Seedlings and Transplanting

Nursery is sown in the month of April to raise the seedlings for planting of seed crop. With 80 percent germination, 600 g seed is required for planting of one-hectare land. Seeds are sown in 10 cm apart and 2 cm deep furrow. Watering and other standard practices are followed for raising healthy seedlings. In agro-climatic conditions of cold desert, 4 to 6 week old seedlings are transplanted in well prepared flat beds at a distance of 60 cm x 45 cm for cabbage and 45 cm x 45 cm for knolkhol. Transplanting should be carried out in the evening and to minimize transplanting shock, the field be irrigated immediately after transplanting.

2.1.2 Field Preparation and Mixing of Manure

Cabbage and knolkhol crops respond well to application of manure and fertilizers due to their heavy nutrient requirement. Fields should be well prepared by ploughing two to three times and incorporating 25 to 30 tones of well decomposed farmyard manure. To obtain good yield, 120 kg nitrogen, 80 kg phosphorus and 75 kg potassium per hectare for cabbage and 100 kg nitrogen, 60 kg phosphorus and 50 kg potassium per hectare for knolkhol are required.

2.1.3 Irrigation

Adequate supply of moisture is required for good crop growth. Depending upon the soil type and weather condition, irrigation should be given at an interval of 8 to 12 days. Irrigation frequency should be reduced when the head/knobs begin to mature and it should be completely stopped at maturity stage. Excess irrigation at maturity stage can cause head splitting in cabbage.

2.1.4 Intercultural Operations

Depending upon weed population, two to three manual weeding should be done. This can be minimized by using plastic/polyethylene mulch which simultaneously also reduces requirement of irrigation water.

2.1.5 Harvesting and Storage of Heads and Knobs

To store the heads of cabbage and knobs of knolkhol, whole plant is uprooted during October to mid November and re-planted in an underground store. In underground store, soil is prepared just like as for transplantation of the crop in the open field. Sandy or sandy loam soil is suitable for re-planting of the plants in underground store. There is no need to add manure and fertilizers as the plants remain in dormant stage during storage. The underground store maintains adequate temperature in winter by harnessing the soil heat and protecting from outside cold. Uprooted whole plants are planted at a distance of 30 cm x 30 cm (close planting) or plant to plant gap may be decreased or increased depending upon the size of heads of cabbage and knobs of knolkhol. To minimize the water loss from the plants through evapo- transpiration, a light watering should be given immediately after re-planting. Thereafter, very light irrigation should be given at an interval of 15 to 20 days; interval of watering may be increased during December-January as slightly higher humidity is observed in these months. In general basement of house is used as underground stores. If independent underground store is made then care should be taken to maintain inside temperature by covering the roof of the underground store with polythene sheet or alfalfa hay. If proper care is taken during the storage, the survival rate of the plants is as high as 80 to 90 percent. Freezing temperature inside store, due to inadequate care will spoil the heads of cabbage and knobs of knolkhol.

2.2 Second Growing Season

2.2.1 Plantation for Seed Production

During the months of March-April when minimum temperature reaches minus 4°C, the dormant plants are uprooted from the underground store and are planted in well prepared field beds. Distance of plants should be followed as per the variety. In case of Drum Head variety of cabbage 75 x 60 cm spacing has been found suitable, however Golden Acre produced maximum seed yield at 60 x 45 cm spacing. 45 x 45 cm spacing is recorded suitable to get maximum seed yield of knolkhol. Standard cultural practices are followed to keep the field free from weeds. Adequate moisture should be maintained by irrigation of the crop at proper interval.

2.2.2 Inspection and Roguing

The crop should be inspected before the marketable stage. Off-type plants should be rogued out on the basis of head shape, colour of leaves at maturity time of the heads/knobs in the first growing season. Minimum three inspections should be made, first before the marketable stage of heads/knobs, second before storage in the underground store of heads/knobs and the third at the flowering stage.

2.2.3 Isolation

Cabbage and knolkhol being highly cross-pollinated crops require isolation from the fields of other varieties and the same variety, which do not confirm to purity. The isolation distances depend upon direction of the crop from field of other crop. Low insect population is found in this region and mostly wind blows from western side to eastern side with high speed therefore, the isolation distance should be more (2000-2500 m) from other varieties in case the seed crop is planted in eastern side of other crop field. But 800-1000 m distance is sufficient when the crop is planted in any other direction of other crops.

2.2.4 Harvesting and Threshing

When the siliquas turn yellow and the seed becomes brown, harvesting can be commenced to avoid shattering. Seed crop may be harvested in two to three lots. Individual plants can be harvested first and the remaining crop can be harvested when about 75% of the pods turn yellowish brown. Harvesting can be done with sickle. Plants are stacked on tarpaulin/clay soil threshing floor, so that shattered seeds are not lost. About 10 to 15 days drying is usually necessary before threshing. Pods have a natural tendency to dehisce and threshing is easy when pods are dry. After threshing, seed should be processed and stored properly. Average 13.15 q/ha seed yield from Drum Head variety, 7.2 q/ha from Golden Acre variety of cabbage and 6.8 q/ha from White Vienna variety of knolkhol has been recorded following two subsequent experimentations. Under ambient conditions the seed can be stored up to 5 years with optimum viability (Table 1,2 and 3).

Table 1 : Performance of cabbage & knol-khol seed crop

Observation (W.V.)	Cabbage (G.A.)	Knol khol
1. Plant height (cm)	109.78	103.84
2. No of branches/plant		
i) Primary	7.3	7.45
ii) Secondary	25.6	21.0
3. No of siliqua/plant	597.75	619.75
4. Length of siliqua (cm)	7.65	7.51
5. Width of siliqua (cm)	.475	7.51
6. No of seeds/siliqua	18.76	19.265
7. Wt of seeds/plant (g)	47.36	50.265
8. 1000-seed wt (g)	5.520	3.875
9. Seed yield (q/ha)	7.2	6.8

Table 2 : Field and Seed Standards (%) for Cabbage Seed Production

Factors	Foundation	Certified
- Off-type (Max.)	0.10	0.50
- Other crop plants (Max.)	–	–
- Objectionable weed plants (Max.)	–	–
- Diseased plants (Max.)	0.10	0.50
SEED STANDARDS		
- Pure seed (Min.)	98.0	98.0
- Inert matter (Max.)	2.0	2.0
- Other crop seeds (Max.)	0.05	0.05
- Total weed seeds (Max.)	0.05	0.05
- Objectionable weed seed (Max.)	–	–
- Germination (Min.)	75.0	75.0
- Moisture (Max.)	7.0	7.0

Table 3 : Field and Seed Standards (%) for Knol khol Seed Production

Factors	Foundation	Certified
- Off-type (Max.)	0.10	0.50
- Other crop plants (Max.)	–	–
- Objectionable weed plants (Max.)	–	–
- Diseased plants (Max.)	0.10	0.50
SEED STANDARDS		
- Pure seed (Min.)	98.0	98.0
- Inert matter (Max.)	2.0	2.0
- Other crop seeds (Max.)	0.05	0.10
- Total weed seeds (Max.)	0.05	0.20
- Objectionable weed seed (Max.)	–	–
- Germination (Min.)	70.0	70.0
- Moisture (Max.)	7.0	7.0

3. Seed Production of Radish, Carrot, Turnip and Beet Root

Radish, carrot, turnip and beet root are major root crops commonly grown in Ladakh. These crops are classified into two distinct groups from seed production point of view viz. annuals or tropical and biennials or temperate. The tropical types can be grown for seed production in the plains but the temperate types produce seeds only in hills/under temperate climate. Temperate type undergoes dormancy or rest period which can only be broken when the roots are subjected to low temperature, i.e less than 6°C for a period of 4 to 6 weeks at any time either in storage or under field conditions. It is for this reason that the seeds of temperate types can only be produced in places like Kashmir, Kullu, Katrain, Kalpa, Darjeeling and Nilgiri hills where winters are severe. Defence Institute of High Altitude Research has identified cold desert areas of Ladakh as potential areas for seed production of root crops (Singh *et al.*, 1999). The technique illustrated below has been standardized by DIHAR after extensive trials on seed production in Leh and Nubra valleys of Ladakh.

3.1 First phase

The temperate varieties of these crops require two growing seasons to produce seeds under the prevailing agro-climatic conditions in Ladakh. In the first growing season (First phase), the seed crop is raised for table purpose. It requires deep, fertile, sandy loam soil. In heavy soils, sowing on ridges is preferred. All these crops require good drainage otherwise the roots are likely to become more branched, misshaped and mal-formed.

In the first phase after the preparation of land, seeds of beet root is sown during May depending upon the weather conditions and those of carrot, radish and turnip are sown in the month of June. In Ladakh, roots of these crops are ready for harvesting in the month of October. Cultivation techniques applied are similar to production of table crop. Matured roots are usually pulled by hand or digged out. True to type roots are selected, leaf tops are cut without damaging the crown leaves.

3.1.1 Storage of roots in pits

After selection and cleaning, roots are stored until the next cropping season. The roots are stored in underground pits of about 6-7 feet deep. The length and width of the pit depends upon quantity of roots to be stored. After filling the pit with roots, 60 cm layer of dry sand is laid on the roots.

Pit storage helps to protect the roots from outside temperature during severe winter which at that time is as low as -30°C. The time temperature observed during winters is usually more than 10°C in lower layer, 4-8⁰C in middle and less than 2°C in upper layer in the pit. Some times temperature goes below 0°C in upper 2-3 feet layer. Due to uneven temperature about 20% spoilage in roots has been observed. To avoid the losses of roots by spoilage Defence Institute of High Altitude Research has designed and constructed underground structures for moderating extreme cold conditions and for storage of roots of these crops during winter (October-March).

3.2 Second phase

3.2.1 Re-plantation of roots for seed production

When the ambient temperature reaches to -2 to - 4⁰C (in the month of April), roots are transplanted in well prepared open fields beds.

3.2.2 Spacing

The optimum plant spacing is a must for proper plant growth harvesting of maximum seed yield. Yield and quality of seeds produced are directly proportional to the growth of mother plant. It is observed that the seeds harvested from the thickly populated crops are undersized, weak and poor in quality. This assumption however, does not hold good for all the crops. In carrot, maximum and better quality seed yield is achieved by close planting of carrot roots (45 cm x 45 cm). It is because the major contribution towards total seed yield is from the first and second order umbels. The quality of seed obtained from the third and fourth order umbels are very poor and less in quantity.

Wider spacing (60 cm x 60 cm) is favored for good quality seed production of radish (Cv Local Red). Plant spacing of 45 cm x 30 cm in turnip (Cv. Purple Top Whit Globe) and 60 cm x 45 cm in beet root (Cv. Detroit Dark Red) gives higher seed yields in Leh valley.

3.2.3 Adding manure & fertilizers in the field

Uninterrupted and rapid growth of root crops produce maximum quantity of high quality seed. Adding manure and fertilizers in the field facilitates such growth. A basal dose of farmyard manure/ compost or other locally available organic manure (30 tones/ha) at the time of field preparation is beneficial. When organic manure is applied at a higher rate, the dose of inorganic fertilizers can be reduced. However, application of 80:50:50/N: P: K per hectare for all the root crops have been found sufficient to obtain optimum plant growth and good seed yield. Complete dose of phosphorous and potassium and half dose of nitrogen should be applied in soil as basal doses before transplanting of the root and remaining nitrogen can be given in 2 equal split doses at 25 and 40 days after transplanting the roots.

3.2.4 Irrigation

Irrigation should be given according to local needs. Timely irrigation is very important for obtaining higher seed yields. According to prevailing weather conditions in Ladakh, irrigation should be given at an interval of 10-12 days.

3.2.5 Isolation

Carrot, radish, turnip and beet root are cross pollinated crops. Hence, proper isolation distance should be maintained to produce genetically good quality seed. The isolation distance for these crops should be maintained as described above for cabbage and knolkhol. Turnip and swede should be isolated from each other and both the crops should also be isolated from Chinese cabbage and mustard.

3.2.6 Selection and roguing

Off type plants on the basis of foliage characteristics should be removed from the field itself before the roots are pulled out in the first phase. Root should be conformed to variety characters such as colour, shape, size, diseased and other undesirable roots should be sorted out. The third roguing is done at the time of early vegetative growth in second phase to discard diseased and undesirable plants and fourth at the onset of flowering. Off-types and very early bolters should be removed from the field.

3.2.7 Harvesting and threshing

Pods of turnip shatter very rapidly when they are fully ripe. Therefore, it is advisable to cut the whole crop when 60-70 percent of the pods turn yellow brown in colour. After 4-5 days it is turned upside down and allowed to cure for another 4-5 days. It is then thrashed with ordinary sticks and lifted with hand shifters.

Unlike turnip, shattering is not a problem in radish and so harvesting can be done when pods are fully ripen and the harvested crop is left in the field for drying and curing.

In carrots, the primary umbels mature first so it would be better that the crop is harvested in two or three pickings. Though shattering is also not a problem in carrot, repeated showers after the umbels ripen, affect viability of the seeds. After proper drying the crop is thrashed.

Table 4 : Field and seed standards for seed production of vegetable root crops

Factor	Carrot		Radish		Turnip		Beet root	
	F	C	F	C	F	C	F	C
Field Standards								
Off. Types% (Max)	0.1	0.2	0.1	0.2	0.1	0.2	0.1	0.2
Diseased plants% (Max)	–	–	0.1	0.5	0.1	0.5	–	–
Seed Standards								
Pure Seed %(Min)	95	95	98	98	98	98	96	96
Inert Matter % (Max.)	5	5	2	2	2	2	4	4
Other Crop Seed (Max) No/Kg	5	10	5	10	5	10	5	10
Weed Seed (Max) No/Kg	10	20	10	20	10	20	05	10
Objectionable Weed Seed (Max)	–	–	–	–	–	–	–	–
Germination (Min.)	60	60	70	70	70	70	60	60
Moisture (Max)	8	8	6	6	6	6	9	9

F = Foundation C = Certified

The seed crops of beet root should be harvested when 80 per cent of the seed/seed balls on a plant get hardened and the base of the inflorescence turn brown. Otherwise, there is possibility of shattering of seeds during harvesting. About 10 days drying is essential for easy removal of seed balls from stalk.

Table 5 : Pereformance of carrot cultivar Early Nantes seed production

Observations	2002	2003	Average
Plant height (cm)	102.89	119.63	111.26
No. of branches/plant			
i) Primary	9.9	8.7	9.3
ii) Secondary	20.1	17.5	18.8
iii) Tertiary	59.8	38.3	49.05
No. of umbels/ plant	50.2	46.4	48.3
No. of umblets/ umbel			
i) Primary umbel	77.38	63.12	70.25
ii) Secondary umbel	64.67	54.23	59.45
iii) Tertiary	51.91	48.30	50.10
No of seeds/umblet			
i) Primary umbel	40.8	47.63	44.21
ii) Secondary umbel	38.12	32.21	35.18
iii) Tertiary umbel	28.6	24.42	26.51
Wt of seeds/umbel (g)			
i) Primary umbel	4.330	4.630	4.480
ii) Secondary umbel	2.572	2.351	2.464
iii) Tertiary umbel	1.095	1.108	1.101
1000 seeds weight (g)	2.460	2.320	2.390
Weight of seeds/plant (g)	82.00	76.420	79.21
Seed yield (q/ha)	11.78	12.57	12.17

Table 6 : Performance of radish & turnip seed production

Observations	Radish (Sindhu Local)			Turnip (PTWG)		
	1998	1999	Average	1998	1999	Average
Plant height (cm)	120.9	127.7	124.3	89.91	107.63	98.77
No of branches/plant						
i) Primary	9.2	7.3	8.25	4.4	5.2	4.8
ii) Secondary	38.7	26.7	32.7	–	–	–
iii) Tertiary	52.1	46.2	49.15	–	–	–
No of siliqua/plant	1709.5	1228.6	1469.05	1122.7	963.8	1043.25
Length of siliqua (cm)	5.69	5.73	5.71	5.86	6.2	6.03
Width of siliqua (cm)	0.71	0.73	0.72	0.35	0.39	0.37
No of seeds/siliqua	5.65	5.75	5.7	19.56	18.36	18.96
1000 seeds weight (g)	10.910	11.087	10.998	1.84	1.93	1.885
Weight of seeds/plant (g)	79.910	64.370	72.14	33.88	29.56	32.665
Seed yield (q/ha)	12.19	14.36	13.275	6.41	5.55	5.98

Table 7 : Performance of beet root seed production

Observations	Beet Roots (DDR)		
	1998	1999	Average
Plant height (cm)	101.32	96.37	98.84
No of branches/plant	8.5	6.4	7.45
No of inflorscence / plant	27.17	30.7	28.62
No of seedlets / infloriscence	10.01	8.26	9.13
No of seeds/seedlets	39.73	37.64	38.685
1000 seed weight (g)	11.540	11.920	11.73
Wt of seeds/plant (g)	101.923	93.270	97.596
Seed yield (q/ha)	8.31	12.17	10.24

In all root crops, grading should be done after proper drying, thrashing and cleaning. Under cold desert conditions, the root crop for seed production is ready for harvesting in the last fortnight of September to first fortnight of October. A well managed seed crop can perform for good seed attributes and seed yield as given in Tables 5-7.

References

Garg, I.D., Chandla, V.K., Paul Khurana, S.M. and Singh, N. (2003). Potato cultivation and health status of the crop in Leh-Ladakh (J&K). *Journal of Indian potato association*, 30 (1-2) : 121-122.

Mathew D., Ahmed, Z. and Singh, N. (2005). Formulation of flowering index, morphological relationship and yield prediction system in true garlic aerial seed bulbing. *Horticulture Science*. 40 (7).2036-2039.

Singh, N.; Singh, B. and Sareen, S.K. (1997). High altitude cold desert beet seed production. Indian Horticulture - Jan-Mar, p 57-58.

Singh, N., Singh, B. and Sareen, S.K. (1999). Swede : A potential crop for Ladakh - Indian Horticulture July - September, p 4-5.

Singh, N. and Singh B. (2000). Effect of bulb size on onion seed yield in Leh (Ladakh) : Paper presented in "National symposium on onion - Garlic production and post harvest management - challenges and strategies, NHRDF, Nashik, Nov'19-21, 2000.

□□□

Chapter-4

Fruit Production in Cold Desert of India

Tsering Stobdan[1], Ashish Yadav[1], Sanjai K Dwivedi[2], Konchok Targais[1], Richa Arora[1] and R.B. Srivastava[1]
Defence Institute of High Altitude Research (DIHAR), DRDO, Leh, Ladakh
[2]CEPTAM, DRDO, Metcalfe House, Delhi-54

1. Introduction

A number of indigenous temperate fruit genotypes are grown either as individual trees or small groups of trees in the cold desert of India. The climatic conditions are highly congenial for the production of quality fruits owing to long fruiting season, clear sky and high diurnal temperature variation during growing season. Apple and apricot are the two major fruit crops of the region. They are marked by great differences in terms of biological and pomological traits. The most common type of propagation except for selected cultivars by vegetative means is through seeds. The region represents a great wealth of apple and apricot biodiversity. Introduction of popular cultivars of apple from outside the region has been witnessed during the last few decades. However, introduction of apricot genotypes at large scale has not been witnessed. Apple and apricot fruits are grown in cold desert of India and are prized for their quality. Historically, the premium quality dried apricots are used as are one of the main trading commodity with neighbouring countries. Among the apricot cultivars grown in the region, Halman and Raktsey Karpo are the most important and popular cultivars. Halman is mainly used for preparing premium quality dried apricot and the kernel is sweet. Raktsey Karpo is desert cultivar with white seed cover in contrast to dark brown colour in most apricot genotypes and the kernel is sweet.

2. Present Scenario

Fruit industry in cold arid region though in infancy stage, is fast expanding in recent years. Apple being the most prevalent fruit crops of the cold desert of Himachal Pradesh; while apricot is the major fruit crop of Ladakh region. To date, the delicious group of apple cultivators has dominated, with a recent trend towards spur type cultivators and colour strains, whereas in apricot, Halman, Raktsey karpo, Shakarpara, Charmagaz and Nari are popular. Almond, walnut, pear, peach, plum, cherry mulberry, grape and seabuckthorn are other important fruits of the region. The export quality fruits produced from this region are excellent in respect of colour, taste, shape and keeping quality. With the increasing global temperature and changing weather scenario, the area under fruits is expanding rapidly with the addition of new areas towards higher altitudes. Despite the biotic, abiotic, ecophysiological and geographical factors, the long day length, high light intensity, low atmospheric humidity, cool night and relatively low insect pest and disease incidence in the region act as favourable factors for growing of fruits.

Land use pattern of Leh, Lahaul and Spiti are presented in Table 1. In Leh District, most of the fruit crop plantation is concentrated in three blocks of Khalsi, Leh and Nubra. Khalsi block accounts for 70 percent of the total fruit production in the district. Apricot and apple are the major fruit crops of the district (Anonymous 2006) while in Lahaul and Spiti, apple is the major fruit crop. Area and fruit

Table 1 : Land utilization pattern in Leh, Lahaul and Spiti (Ha) (Anonymous 2006, DOA, 2009)

Particular	Leh	Lahaul	Spiti
Total geographical area	4500000	201987	710111
Forests	2900	134887	482
Cultivated land	10585	2156	887
Area under fruit	1298.4	166.1	223.2

production in Leh and Lahaul & Spiti districts are presented in Table 2 and 3. Apple cultivation is gaining momentum in recent years. Plantation of high yielding apple cultivars is expanding especially in Leh and Spiti districts.

Table 2 : Area and production of fruit in Leh district (Anonymous 2006)

Fruit Crop	Number of Trees (number)	Area (ha)	Est. Production (MT)
Apricot	1,78,140	707	2956.1
Apple	1,43,260	532	3605.1
Pear	568	2.25	7.8
Peach	1030	4.15	6.6
Plum	56	0.3	0.4
Grapes	1105	2.0	9.0

Table 3 : Area and production of fruit in Lahaul and Spiti (DOA, 2009)

Fruit crop	Lahaul		Spiti	
	Area (Ha)	Production (MT)	Area (Ha)	Production (MT)
Apple	166.0	1303.6	197.2	326.2
Apricot	1.0	-	1.0	-
Dry fruits	-	-	0.6	-

3. Standard Propagation Techniques

Vegetative propagation techniques in apricot have been standardized on seedling rootstock under cold arid region. Chip budding (Dwivedi and Singh, 2001a), T-budding and grafting (Dwivedi *et al.,* 2000) are successful at different periods. However, chip budding was found to be the best method (Dwivedi, *et al.,* 2002b). Old seedling apricot plants can be converted to productive plants by Top Working using the shield method of budding during mid June-early July. This has shown 90.0 percent success (Dwivedi and Singh, 1999). For raising root stock, seeds are soaked on water for 8-10 days in February. Stratification is done by keeping seeds between layers of sand in boxes under shade and moist condition. Dormancy breaks after 25-30 days and the seeds are sown in nursery at spacing of 30×15 cm in 1st and 2nd week of April. Light irrigation is required till germination. Under Ladakh conditions 70-80% stratified seeds germinate and 55-60% attain graftable diameter of pencil thickness in November.

Raising of apple nursery has been a major problem in cold deserts. Experiments conducted at the Regional Horticultural Research Sub-Station, Tabo, Spiti have shown that fruit nursery growing is possible under greenhouse conditions. The graftable seedlings procured in February from wet temperate zone are planted in the greenhouse. The seedlings start sprouting in April and grafting is performed. Success rate of tongue grafting has been found better. The plants are ready for sale by the end of winter season. Greenhouses of 10×4×2 m size provide better result than larger greenhouses. The ventilators and doors of the greenhouses need to be kept open during daytime (Dhiman *et al.*, 2009).

Seabuckthorn the golden bush of cold desert Ladakh can be propagated successfully by seed which remains viable for more than three years. The month of May is the best time for sowing of seeds in nursery beds. Before sowing the mature seeds should be stratified for about 25 days by keeping in a box in alternate layers of moist sand. These boxes are kept under cool and shady condition for about 25 days. This method gives over 85 percent germination (Dwivedi *et al.*, 2007b). Seabuckthorn has profuse suckering habit and has been reported to yield 13-40 suckers per plant. These suckers are true to the type that can be separated along with the roots from the mother plants and can be planted directly. The best time for separation of suckers and planting is in March before bud break (Dwivedi *et al.*, 2006). However, the major disadvantage is difficulty in operation due to thorny plant structure. Seabuckthorn can also be propagated successfully by stem cuttings, and so this method has been recommended for commercial propagation. One to two year old stem cuttings having pencil thickness are ideal for its propagation. The stem cuttings are taken from identified male and female plant to get elite planting material. Seabuckthorn cuttings are not easy to root on their own, thus indole butyric acid (IBA) is applied @ 500 ppm before planting in the nursery beds. The best time for planting of cuttings has been found to be the month of April (Dwivedi *et al.*, 2002a). About 85 percent success rate has been recorded in rooting of stem cuttings under Ladakh conditions. These cuttings are ready for plantation in the main field in March of the following year. (Dwivedi *et al.*, 2006). Defence Institute of High Altitude Research has successfully standardized asexual method of propagation of seabuckthorn through soft wood cuttings using 200-300 ppm NAA. The technique has yielded over 90 percent rootability in cuttings with excellent root quality under mist condition having course, sand rooting media. The plantlets get ready for transplantation in poly bags within 90-100 days after planting of cuttings (Dwivedi *et al.*, 2006).

4. Marketing

Marketing of fresh fruits produced in Leh and Kargil is confined to Ladakh region due to restriction on movement of fresh fruit outside the region. However, dried apricots are being sold within and outside the region. Estimate marketed surplus and marketing pattern in Lahaul and Spiti are shown in Table 4. Regional consumption is low and majority of the produce are being sold to distant markets within and outside the State.

Table 4 : Marketing of fruits from Lahaul and Spiti (MT) (DOA, 2009)

	Total production	Consumption	Local market	Marketed Distant market within State	Distant market outside State
Lahaul	1303.3	8.6	153.9	961.8	166.7
Spiti	325.8	4.3	109.5	172.1	31.3

5. Major Insect Pest and Diseases

Incidence of insect pests and diseases in the cold arid region is low due to low temperature and very low relative humidity. However, incidence of codling moth, leaf curling aphid and gummosis are major problem.

5.1 Codling Moth

Codling moth (*Cydia pomonella*) is a serious insect-pest of apple and apricot in Ladakh. The pest is widely distributed in fruit growing areas of Ladakh. To avoid and restrain its spread to other parts of the country the Government has imposed restrictions under SRO 397 dated 8th September 1981 under the Jammu and Kashmir Plant and Disease Act 1973, on the movement of apple and apricot from Ladakh (Dwivedi *et al.*, 2009). Codling moth has a remarkable ability to adapt to a wide range of climatic conditions and geographical locations. The moth passes the winter as full grown larvae in diapauses inside a thick silken cocoon. They usually spin their cocoon under loose bark on the trunks of apple trees, under shelters about the base of the trees, or on the ground nearby. They remain dormant through the winter and are able to withstand severe and freezing temperatures. In mid spring, the larvae change in to pupal stage and adult moths emerge after two to four weeks. In Ladakh, codling moth has well synchronized its life cycle as per the prevailing conditions of the area.

Codling moth damages fruits by stings where larvae bore a short distance in to the fruit. The deep entries occur when larvae penetrate the fruit skin, bore into the core and feed on seed cavity. Larvae may enter the sides, stem end or calyx end of the fruit. Entries into the calyx are often difficult to detect without cutting into the fruit. It is reported that the infestation of codling moth in apple was 49.7 percent and 42.5 percent in Kargil and Leh districts, respectively. It is estimated that about 60 to 70 percent loss is being observed in apple industry each year (Akbar *et al.*, 2010).

Management of codling moth needs a sustainable approach of integrated pest management. An integrated management schedule involves biological control such as *Trichogramma embyrophagum*; installation of pheromone trap; spraying of chemical insecticides such as Cypermethrin, Quinalphos and Chloropyriphos; bur lapping with corrugated gunny bags in two to three layers from base to end of the tree trunk; sanitation of the orchards that include burning of the fallen leaves, twigs and pruned wood, burying of the damaged fruits, scrapping of the bark, manual destruction of the larvae. Involving local community for implementing the programme to eradicate this noxious pest is a must.

5.2 Leaf Curling Aphids

Leaf curling aphid (*Brachycaudis helichrysi*) is a regular pest of peach, apricot, almond and plum in order of preference. It damages floral and vegetative bud by sucking sap due to which unfolding leaves curl up, remain smaller, get distorted and later turn pale. Consequently fruit-set is poor with a pre-mature fruit drop and sub-normal fruits are formed. The badly curled leaf whorls usually dry and shed (Chadha, 2001). It can be managed by spraying of Monocrotophos (40 EC) 200ml in 250 ltrs of water at 10-15 days interval (Dwivedi *et al.*, 2007a).

5.3 Gummosis

Gummosis is a major problem of stone fruits in cold desert condition. Exudation of gum from the stem and branch loosen the bark causing secondary infection leading to drying of branches in severe cases. The causal factor could be biotic such as fungi and bacteria (*Pseudomonas syringae*) or may be due to abiotic factors. For management of the disease, scarification of gumming area with sharp knife in spring followed by application of Mashobra paste is effective. Foliar spray of streptocycline (0.01%) plus copper oxychloride (0.2%) is effective to overcome gummosis (Dwivedi *et al.*, 2007a).

6. Interventions Required for Fruit Crop Development

For development of fruit cropping in cold desert of India, integrated compact area development approach need to be followed. Need-based support should be provided for crop production with forward and backward linkages. Inputs such as supply of quality planting material of only recommended varieties, post-harvest infrastructural facilities and processing units need to be established. Available arable land need to be utilized by changing crop priority. Wasteland need to be promoted for growing horticultural crops. Cost-effective polyhouses need to be promoted in the region. Skilled manpower at different levels is required. There is a need to promote integrated nutrient management through vermi-composting, biofertilizer and judicious use of chemical fertilizer. Farm mechanisation conducive to hill farming is required to reduce labour use. Human resource need to be developed and training programmes on modern aspects of fruit crop production and post-harvest management need to be organised. Setting up of large demonstration blocks should be done on priority. Development of integrated disease and pest management modules suitable for organic fruit production is required. Hi-tech horticultural production technology such as microirrigation, protected cultivation, micropropagation etc. need to be introduced in the region. Farmer advisory systems to address their day-to-day queries need to be created in every block. Studies on economic and marketing aspects of fruit crops need to be conducted.

7. Need for Promotion of Processing Units

In view of restrictions on the movement of apple and apricot from Ladakh, special emphasis should be given to development of processing units in the region. As of date there are only four processing units in Kargil district and 11 in Leh district. Nine units have FPO licence under Home scale while four have Cottage scale licence and two units have Small scale licence. Most of the units remain functional for few months and production activity remain restricted to primary processing of seabuckthorn berries and production of jam and squash. The finished products are mainly sold in local market. Besides establishment of new processing units, the processing capacity of existing units needs to be augmented and existing facilities need to be modernised. Human resource need to be developed and prescribed international and domestic standards should be disseminated and adhered to. Food processing industries need to be adopted and special package for its promotion need to be initiated. Unless the scheme on integrated development of fruit crops takes into consideration the need of processing units in the region, balanced regional growth cannot be promoted.

8. SWOT Analysis for Improving Fruit Production

8.1 Strength

i. The cold desert region is endowed with agro-climatic conditions that are conducive to grow several high value fruit crops. The best apples are being produced in cold desert region.

ii. There are vast tracts of cultivable waste land that could be brought under cultivation.

iii. There are plenty of water resources such as glaciers, streams and rivers that have not been judiciously used in the region. If harnessed judiciously, the water resources can substantially augment area under fruit crops.

iv. Very low amount of chemical fertilizers and pesticides are being used in the region and thus organic fruit can be promoted and marketed.

v. Incidence of insect pests and diseases are less in cold arid region.

vi. Unique climatic conditions provide opportunity to produce high quality fruit crops.

vii. Research organisations like Defence Institute of High Altitude Research, Regional Agricultural Research Stations of Dr Y S Parmar University of Horticulture and Forestry, CSK Himachal Pradesh Agricultural University, Sher-e-Kashmir University of Agricultural Science and Technology have been established in the cold deserts of Jammu & Kashmir and Himachal Pradesh to focus on location specific research. It is expected that sub-stations of other National Research Organisations are soon going to be opened up in Ladakh region.

8.2 Weakness

i. The cold desert regions are inaccessibe especially during winter months. The villages scattered over a vast area are not well connected.

ii. The harsh climatic conditions limit the cropping season to one crop season only.

iii. Due to low precipitation, cultivation is confined to areas that can be irrigated.

iv. Damage caused by spring frost is a major problem in cold desert region.

v. Restrictions on the movement of apple and apricot from Ladakh are a major hindrance for development of fruit crops in Ladakh region.

vi. Weak processing infrastructure.

vii. Unorganised plantation of fruit crops.

8.3 Opportunities

i. As a result of rising temperature and improved agro-practices, the areas in cold desert that are not suitable for fruit crops few decades back are emerging as the potential areas to grow fruit crops like apple and apricot.

ii. Seabuckthorn, an ideal plant for cold desert, can be grown on large scale for income generation and ecological benefits. The resource has not been trapped fully all over the region.

iii. New fruit crops are being introduced in the region by several R&D organisations and Universities. This will broaden the fruit base of the region.

iv. Organic fruit production and wild harvest of seabuckthorn can be promoted to fetch higher price for the fruits grown in cold desert region.

8.4 Threats

i. The ongoing process of globalisation poses threat to cultivation of fruit crops. Farmers may get lesser price for their produce as a result of cheaper imports from other Asian and European countries.

ii. Due to mountainous topography and unexpected heavy snowfall, there is a limited time period for supply of critical inputs.

9. Future Prospects

Fruit production in cold desert is in infancy stage. There is a wide gap between the existing production and potential production of fruit. It has been estimated that after realizing the full potential, apple production in Lahaul and Spiti would increase from existing 1,629 tonnes to about 13,024 tonnes showing around 70% growth of which major increase will be in Spiti (DOA, 2009). Similarly, there is immense scope for production of apple and apricot in Leh and Kargil districts. Apart from apple and apricot, other fruit crops like peach, plum, cherry, almond, pear and strawberry are showing promising results under field trial. These fruits trees can increase the fruit base of the cold desert of India.

During the last five years there has been a major interest on seabuckthorn. Mega projects on seabuckthorn have been initiated by several R&D organizations in view of its biotechnological, neutraceuticals, pharmaceutical and socio-economic potential (Stobdan *et al.*, 2008). Seabuckthorn has now been included under the Green India Mission. Large scale plantation of seabuckthorn will generate income for the people of cold arid region besides ecological benefits (Stobdan *et al.*, 2011).

References

Akbar, P.I., Jahangeer, B.A., Malik, A.R. and Parray, G.A. (2010). Codling Moth (*Cydia pomonella*) pest in cold arid Ladakh: Problems and Management. http://krishisewa.com/articles/2010/codlingmoth.html

Anonymous. (2006). *Economic Review of Leh District* (2005-2006). Ladakh Autonomous Hill Development Council, Leh Ladakh.

Chadha, T.R. (2001). *Text Book of Temperate Fruits*. Indian Council of Agricultural Research, New Delhi, India

DOA, (2009). District Agriculture Plan: Lahaul-Spiti, H.P., Vol. VII.

Dhiman, K.R., Bhardawaj, D.R., Bawa, R. and Vikram A. (2009). Producing world class apples in cold deserts for health and economic security- a case study of Regional Horticultural Research Sub-Station, Tabo in Himachal Pradesh. In *Advances in Agriculture, Environment & Health*. (eds. S.B. Singh, O.P. Chaurasia, A. Yadav, A.M. Rimando, T.H. Terrill). SSPH Delhi, India. pp. 99-104.

Dwivedi, S.K. and Singh, B. (1999). Topworking of apricot in Ladakh. *Progressive Horticulture*, 31 (1-2): 29-31.

Dwivedi, S.K., Singh, B. and Paljor, E. (2000). Studies on vegetative propagation of apricot through grafting in Ladakh. *Indian Journal of Horticulture*, 57(1): 39-41.

Dwivedi, S.K. and Singh, B. (2001a). Studies on chip budding in apricot (*Prunus armeniaca*) arid conditions of Ladakh. *Indian Journal of Horticulture*, 58(3): 215-217.

Dwivedi, S.K., Singh, B. and Attrey, D.P. (2001b). Studies on distribution, propagation and ultilisation of seabuckthorn in Ladakh. Seabuckthorn-a resource for health and environment in 21st century. In *Proceedings of International Workshop on Seabuckthorn*, Feb. 18-21, 2001. (eds. V. Singh and P.K. Khosala), CSKHPKV-Palampur, India. pp. 20-24.

Dwivedi, S.K., Attrey, D.P. and Kareem, A. (2002a). Post harvest management of seabuckthorn (*Hippophae* spp.) in cold arid regions of Ladakh. In *National Workshop on Post-harvest Management of Horticultural Produce*, 7-8 Feb. 2002. National Productivity Council, India. pp. 89-98.

Dwivedi, S.K., Attrey, D.P., Paljor, E. and Singh, B. (2002b). Studies on conservation of genetic diversity in apricot in cold arid region of India. *In* 26th International Horticultural Congress, Toronto (Canada), 11-17 August, 2002. pp. 524-25.

Dwivedi, S.K., Singh, R. and Ahmed, Z. (2006). *The Seabuckthorn*. Field Research Laboratory, DRDO, Leh, Ladakh (J&K), India.

Dwivedi, S.K, Kareem A. and Ahmed, Z. (2007a). *Apricot in Ladakh*. Field Research Laboratory, Leh Ladakh, India

Dwivedi, S.K., Hemraj and Ahmed, Z. (2007b). Standardisation of a fast method of vegetative propagation of Seabuckthorn through soft wood cutting in India. In *3 rd International Seabuckthorn Association Conference*, Quebec, Canada, August,12-16, 2007.Paper No83.

Dwivedi, S.K., Stobdan, T., Yadav, A., Singh, S.B. (2009). Fruit production in cold arid regions of India: challenges and opportunities. In *Advances in Agriculture, Environment & Health*. (eds. S.B. Singh, O.P. Chaurasia, A. Yadav, A.M. Rimando, T.H. Terrill). SSPH Delhi, India. pp. 61-81.

Stobdan, T., Angchok, D. and Singh, S.B. (2008). Seabuckthorn: An emerging storehouse for researchers in India. *Current Science*, 94 (10): 1236-1237.

Stobdan, T., Chaurasia, O.P., Korekar, G., Yadav, A. and Singh, S.B. (2011). Scope of Seabuckthorn (*Hippophae rhamnoides* L.): A Wild Shrub for Sustainable Livelihood in Cold Desert of Ladakh, India. In *Global Change, Biodiversity and Livelihoods in Cold Desert Region of Asia*. (eds. K.G. Saxena, L. Liang and X. Xue). BSMS, Dehradun, India. pp. 255-258.

□□□

Chapter-5

Seabuckthorn Production for Greening and Sustainable Income Generation in Cold Desert of India

Tsering Stobdan[1], Girish Korekar[1], O.P. Chaurasia[1], B. Balaji[2], Ashish Yadav[1], Sanjai K. Dwivedi[3], Konchok Targais[1], Sunil Mundra[1] and R.B. Srivastava[1]

Defence Institute of High Altitude Research (DIHAR), DRDO, Leh, Ladakh

[2]Divisional Forest Department, Leh Ladakh

[3]CEPTAM, DRDO, Metcalfe House, Delhi-54

1. Introduction

The actinorhizal plant seabuckthorn (*Hippophae rhamnoides* L.) of the family Elaegnaceae has the unique characteristics to grow in marginally fertile soil of cold desert. The shrub can withstand extreme temperature from -43°C to +40°C and is considered drought resistant (Li, 2002). These two characteristics make the shrub an ideal plant species for cold deserts.

Seabuckthorn berries are among the most nutritious fruits. Concentration of pro-vitamins A, B_2 and C of the berries is much higher than other fruits and vegetables such as orange, carrot and tomato. Thus, the berries are a storehouse of vitamins in the region where availability of other vitamin rich fruits are limited. Seabuckthorn berries have a unique characteristic of remaining intact on the shrub throughout the winter months despite the subzero temperature. Many bird species feed on the berries when other sources of food are limited in the region. Seabuckthorn finds mention in the writings of ancient Greek scholars such as Theophrastus and Dioscorides. The plant is known as a remedy for maladies of horses. Leaves and branches were added to fodder to induce rapid weight gain and shiny coat, and in fact, the generic name *Hippophae* is classical Latin for 'shinning horse' (Lu, 1992). The leaves serve as protein rich fodder for cold desert animals like sheep, goat, donkey, cattle, double humped camel etc. The thorny and bushy growth of the shrub provides a protective shelter for flora and fauna thereby maintaining the fragile ecosystem of the cold arid region.

2. Geographical Distribution

Hippophae rhamnoides, H. salicifolia and *H. tibetana* are the predominant seabuckthorn species in India. Of these, *H. rhamnoides* is widely distributed in the country. Seabuckthorn (*H. rhamnoides*) is naturally distributed in six valleys namely Leh, Nubra, Changthang, Suru, Lahaul and Spiti in Trans-Himalayan cold deserts of India. It is also found in few pockets of Nathula in Sikkim. Remote sensing and Geographical Information System (GIS) have been used to map seabuckthorn in Lahaul and Spiti in Himachal Pradesh (Roy *et al.*, 2001) and Ladakh region in Jammu and Kashmir (Dwivedi *et al.*, 2006). The high altitude cloudless skies, arid and sparse vegetation cover provides an ideal condition for mapping in the hostile and inaccessible terrains. Studies on satellite images by DIHAR and Defence Electronic Application Laboratory, DRDO, have shown that 11,500 ha of land is under pure natural seabuckthorn in Ladakh. The shrub is mainly distributed in Nubra, Indus and Suru valleys. It is also found in Changthang valley at an altitude of 15,400 ft above MSL. Natural plantation is mainly distributed along Nubra, Shyok and Indus River. Plantation is done around agricultural fields to serve as wind break and biofence.

3. Ecological Benefits

Seabuckthorn brings many environmental benefits, including soil and water conservation, desertification control, land reclamation, erosion and water loss control, reforestation, and the establishment of wild life habitats especially in fragile ecosystem due to its extensive root system coupled with efficient nitrogen fixation. In 10-12 year old plant, roots extend up to 537 cm horizontally and up to 127 cm vertically, making it ideal for soil binding. The shrub has been found growing well in sandy, rocky, saline and ravine soils. Areas receiving 400 mm of annual precipitation is suitable for seabuckthorn orchard planting. Because of its resistance to drought and ability to tolerate soil salinity and low temperatures, it is suitable for many conditions that are simply too demanding for other plants. The shrub is among the first plant species that can grow in any barren land. Establishment of seabuckthorn plantation led to improved soil fertility and moisture conservation leading to shrub-grass community of thick forest after 4-5 years. Roots of seabuckthorn have symbiotic association with bacterium belonging to genus *Frankia*. The symbiotic association fixes atmospheric nitrogen in soil thereby increasing soil fertility of nutrient deficit soil in deserts. It is estimated that seabuckthorn plantation fix 180 kg of nitrogen per hectare every year (Jike and Xiaoming, 1992). In India, the Jammu & Kashmir State Forest Department has taken initiative in controlling soil erosion and land reclamation through plantation of seabuckthorn (Stobdan *et al.*, 2008b). Traditionally, the shrub is planted along river sides and canals to prevent soil erosion. Thorny and bushy growth of seabuckthorn provides a protective shelter for flora and fauna thereby maintaining the fragile ecosystem of the cold arid region. In view of the immense ecological benefits of seabuckthorn, large scale plantation of the shrub has been included under the Green India Mission by Ministry of Environment and Forest, Government of India.

4. Traditional Uses

Seabuckthorn has been judiciously used by people living in the cold deserts. Due to scarcity of resources, seabuckthorn has been used traditionally for a variety of purposes. Every part of the plant viz. fruit, leaf, twig, root and thorns have been traditionally used as medicine, nutritional supplement, fuel and fence and therefore, seabuckthorn is popularly known as 'Wonder Plant', 'Ladakh Gold', ''Golden Bush' or 'Gold Mine' of cold deserts. The agricultural fields are valued the most in the region due to limited cultivable land. Traditionally the dense and thorny shrub is planted around agricultural field and plantation sites to protect against stray animals and pedestrian movement. The dried twigs and branches are also put along the boundary walls of residential houses and fields. Plantation of timber trees like willow and poplar is an important activity in cold desert of Ladakh. The cuttings during its initial years need utmost care especially from the stray animals. To prevent animals from damaging the plants, seabuckthorn branches are tied around the cuttings to serve as tree guard. This practice is effective in increasing survivability percentage of the plants during early stages. Cold deserts are characterized by high wind velocity leading to environmental degradation. Windbreaks made of seabuckthorn are effective at preventing wind erosion in open areas. The shrub can resist drying effect and physical injuries caused by wind. The extensive root system of the shrub is ideal for soil binding. Due to extensive root system, seabuckthorn plantation is being done around the water channel to check erosion due to water flow.

Cold deserts have a typical problem of firewood. On one hand, the region is cold and thus requires fuel wood to keep the houses warm during winter months. However, at the same time due to low forest cover, availability of wood for firewood purpose is a major challenge. Under such circumstances, seabuckthorn stem and branches are being used as firewood. The calorific value of dry seabuckthorn is 4,785.5 calories per kg. The shrub is fast growing and can be stumped after every 3-5 years. Since the shrub grows fast and tolerates repeated cuttings, it reduces the harvesting pressure on other native woody plant species like poplar, willow and juniper. Six-year old seabuckthorn plantation on one hectare can produce 18 tons of firewood which is equal to nearly 12.6 tons of standard coal. The firewood is poplar especially in monasteries and during ceremonies where the requirement of high calorific firewood is required. Most monasteries maintain its own seabuckthorn growing areas to ensure regular supply of firewood. Seabuckthorn thus proved to be a popular green plant in the region. Charcoal that remains after burning seabuckthorn is being used by blacksmith. Seabuckthorn stem is hard and often used as handle for agricultural implements.

Seabuckthorn has been used in traditional 'Amchi' system of medicine for centuries in Ladakh. The medicinal value of seabuckthorn was recorded as early as the 8th century in the Tibetan medicinal classic *rGyud Bzi* (Four Text of Fundamental Tibetan Medicine). More than 200 seabuckthorn-based formulations are being traditionally used either in pure form or mixed with other medicinal plants. The most popular formulations are being used in treatment of lung and phlegm disorders, blood disorder, menstrual problem, throat infection, liver problem, spleen and stomach disorder, cancer and diabetics (Gurmet, 2009). Modern research has supported the medicinal properties of seabuckthorn.

5. Nutritional Value

An interesting fact of seabuckthorn is that there are separate male and female plants. Only the female plants bear yellow, orange or red berries that are ripen in August-September. The berries are among the most nutritious fruits. Vitamin C content ranges from 200-2500 mg/ 100g berries, which is higher than many other vitamin rich fruits like amla, orange, kiwi etc. Results of analysis of seabuckthorn berry from Leh valley of Trans-Himalayas showed the presence of high content of multivitamins including vitamin C (275 mg/100g), vitamin A (432.4 IU/100g), vitamin E (3.54 mg/100g), Riboflavin (1.45 mg/100g), Niacin (68.4 mg/100g), Pantothenic acid (0.85 mg/100g), vitamin B-6 (1.12 mg/100g), and vitamin B-2 (5.4 mcg/100g). Similarly mineral element composition revealed high amount of minerals including potassium (647.2 mg/l), calcium (176.6 mg/l), iron (30.9 mg/l), magnesium (22.5 mg/l), phosphorous (84.2 mg/l), sodium (414.2 mg/l), zinc (1.4 mg/l), copper (0.7 mg/l), manganese (1.06 mg/l) and selenium (0.53 mg/l) (Stobdan *et al.*, 2010). Thus the berries are storehouse of vitamins in the region where availability of other vitamin rich fruits is limited.

6. Genetic Diversity

Seabuckthorn is a cross-pollinated crop and hence there exists high level of genetic diversity. Studies conducted by DIHAR (DRDO) on 17 morphotypes of seabuckthorn collected from Nubra valley revealed large genetic diversity. RAPD markers show five major groups in the natural population. None of the morphotypes shares a similarity more than 25 percent indicating high level of genetic diversity between the morphotypes (Singh *et al.*, 2006). Extensive survey in Ladakh region has shown presence of more than 200 genotypes in the natural population with wide variation in physico-chemical properties. It has been recorded that the shrub height varied from 6-24 ft. Inverse relation has been found between the shrub height and the altitude of the region in which the shrub grows (Dwivedi *et al.*, 2006). Morphological studies in natural population of seabuckthorn in Lahaul, Spiti and Kinnaur has also shown a similar trend. Plant height is greatest in Kinnaur and lowest in Spiti (Sankhyan and Sehgal, 2005). Based on physico-chemical properties of the berries and the leaves, 19 morphotypes with distinct variation are being maintained at DIHAR, Leh. A set of morphological descriptors for the genus *Hippophae* has been developed for germplasm obtained from various parts of India. The pattern of variability in morphological characteristics associated with habitat, plant growth habit, stem, leaf, fruit, flower, seed and biochemical parameters are taken into consideration. The suitability of these parameters in characterization of *Hippophae* germplasm has been successfully demonstrated. Using the descriptors, it is found that *H. rhamnoides* and *H. salicifolia* are genetically closer than that with *H. tibetana* (Mathew *et al.*, 2007). Two hundred accessions of seabuckthorn have been collected from Leh and Nubra valley and maintained at field gene bank of Defence Institute of High Altitude Research. The accessions were characterised for morphological parameters, nutritional content and genetic diversity by Amplified Fragment Length Polymorphism (AFLP) markers.

7. Morpho-Biochemical Variations

Among the three species, *H. rhamnoides* is a shrub or small tree form (2-5 m), *H. salicifolia* is a small or large tree (4-9 m), whereas *H. tibetana* is a under shrub (0.2-0.6 m). Fruit size varies from 8 to 27 g/100 fruits in *H. rhamnoides,* 30-35 g in *H. salicifolia* and 15-21 g in *H. tibetana.* Colour of the fruit varies from red to reddish orange in *H. rhamnoides,* yellow in *H. salicifolia* and orange with green strips in *H. tibetana.* Seed size also varied from 0.718 to 1.103 g/100 seeds in *H. rhamnoides,* from 0.943 to 1.142 g in *H. salicifolia* and from 1.050 to 1.331 g in *H. tibetana.* The seed colour is dull white in *H. tibetana,* whereas it varied from light to dark brown in other two species. The shape of the seeds were oblong to ovate, oval, elongate and elliptical in *H. rhamnoides,* ovate in *H. salicifolia* and elliptical in *H. tibetana.* The moisture content in the fresh fruits varied from 72.9 to 85.0 percent. The total oil content varied from 3.2 to 4.6 percent in *H. rhamnoides* and 2.0 per cent in *H. salicifolia.* The vitamin C content varied from 219-642 mg/100g in *H. rhamnoides* to 947 mg in *H. salicifolia* and 356 mg in *H. tibetana.*

Among the 12 biotypes of seabuckthorn growing in different areas of Lahaul-Spiti, weight of 100 seeds varied from a maximum of 1.331 g in *H. tibetana,* to 1.142 g in *H. salicifolia* and 1.103-0.718 g in *H. rhamnoides.* The total oil content was maximum in the seeds of *H. tibetana* (16.9%), which was significantly higher ($p<0.05$) than the seeds of *H. rhamnoides* (8.8-14.6%) and *H. salicifolia* (10.5%). There was no definite pattern of different characteristics of the seeds of different biotypes of seabuckthorn in relation to altitude of the region (Singh *et al.,* 2005).

From the different regions of Ladakh, morpho-biochemical variation in the ripe fruits were also studied by scientists of DIHAR, Leh (Dwivedi *et al.,* 2005, 2006). The colour of ripe fruits varied from pale-yellow, deep-yellow, orange-yellow, greenish-yellow, reddish-yellow, yellow and light red. The shape of ripe fruits have also shown considerable variation as round, oval, oblong, ovate, elongate, elliptical and even capsular shape with flat and pointed end. The average fruit length varied from 5.0-11.1 mm. Similarly, fruit width ranged from 4.2-7.8 mm. Minimum weight of 100 ripe fruits was observed to be 7.4 g while maximum weight was 24.7 g. The juice content varied from 56.4 percent to 84.0 percent in ripe fruits of the region. The fruits contain 2.0-12.5 per cent seed. Fruit juice contains 6.2-18.2°Brix TSS. The total sugar varied from 3.45-6.1 percent. The vitamin C content ranges from 88-1422 mg/100g of ripe fruits. Seeds of seabuckthorn have shown variation in colour, light-dark brown to blackish brown and dull brown. Shape varied from oval to ovoid and oblong, elongate to elliptical. Weight of 100 seeds is about 1.24 g while length of seed was 4.4 mm and width is 2.5 mm. The seed contained 9.0-16.9 per cent valuable oil (Dwivedi *et al.*, 2005).

Morphological changes during ripening of the fruits was studied in a natural stand of seabuckthorn growing at 2650 above MSL, on the riverside in Lahaul. It was observed that fruit size varied from 8.2-16.0 g/100 fruits with a maximum value of 17.7 g during the 4th collection on 8th September. Vertical diameter varied from 5.4 to 7.8 mm with a maximum of 8.0 mm. Cross diameter varied from 5.5 to 6.2 mm with a maximum value of 6.9 mm. The length of fruit stalk varied from 2.9 to 3.2 mm. Colour of the fruit was initially light yellowish green with both ends light red. After one month, fruits were yellow with both ends were red. After 2 months, fruit were reddish yellow. Initially fruits did not break easily during plucking. After one month, fruit skin damaged, if not collected with extra care. After two months, fruit released the reddish white paste like material during plucking. The total soluble solid varied from 8.7 to 13.8 per cent.

Kala *et al.* (2003) working at Regional Research Laboratory of CSIR at Thiruvananthapuram, established a protocol for the isolation of phytochemicals of the fruits of seabuckthorn growing in Ladakh. Carotenoids, tocopherols, fatty acids and sterols were obtained from the non-polar fractions. Terpenoids were mainly isolated from the non-saponifiable part of the seed oil. Phenolics and carbohydrates were obtained from the polar fractions in addition to xanthophylls, beta-carotene was the major carotenoid in fruit pulp and seeds. Sitosterol was identified as major steroid. High content of Ursolic acid in the seeds and leaves. Beta-Amyrin and Uvaol were found as minor terpenoids. Among the fatty acids, were present Palmitic, Palmitoleic, Stearic, Oleic, Linoleic and Linolenic acids. The polar fractions have complex glycosides of flavonboids, i.e., isorhamnetin and kaempferol.

The mineral composition (Zn, Cu, Fe, Ca, Mg, Na and K) of seabuckthorn fruit pulp and seed components were studied by Singh *et al.* (2001) in 8 populations of seabuckthorn (*H. rhamnoides* ssp. *turkestanica* and *H. salicifolia*) growing in Lahaul valley (2600-3200 AMSL), a dry temperate Himalayas region. In the fruit pulp of *H. rhamnoides*, Zn varied from 8.7 to 12.5 ppm; Cu, 3.1-4.9 ppm; Fe, 427-800 ppm; Ca, 2310-3325 ppm; Mg, 500-585 ppm; Na, 55-137 ppm and K, 8575-10025 ppm and in seed, Zn varied from 20.7 to 27.8 ppm; Cu, 4.1-8.4 ppm; Fe, 18.1-74.7 ppm; Ca, 480-1380 ppm; Mg, 1080-1725 ppm; Na, 60-133 ppm and K, 6225-9350 ppm. The values of Zn, Cu, and Mg were higher in the seeds as compared to pulp, where as other minerals (Fe, Ca and K) were higher in pulp, with the exception of Na, which was almost similar in both parts. The values for *H. salicifolia* were in general in the range of *H. rhamnoides* populations.

8. Product Development and Income Generation

Seabuckthorn has immense potential to uplift socio-economic condition of the local populace. Studies conducted using satellite imagery by Defence Institute of High Altitude Research and Defence Electronic Application Laboratory has shown that about 11,500 ha of land is under seabuckthorn in Ladakh region (Dwivedi *et al.*, 2006). Until August 2001, seabuckthorn was considered a thorny menace by locals due to its profuse suckering nature. The plants are being uprooted due to fear of its spread to fertile soil. In India, seabuckthorn has received increasing attention after Defence Institute of High Altitude Research (DRDO) has developed the technology for preparing beverage from its highly acidic fruit (Stobdan and Singh, 2009). The patented technology has been transferred to commercial units, NGOs and local entrepreneurs. The technology is well received by industrialists and ready-to-serve beverage is currently available in Indian market under brand name of 'Leh Berry', 'Ladakh Berry', 'Power Berry' *etc.* (Stobdan et al., 2009). Tender and healthy leaves from male plant are mixed with other high altitude herbs like salam panja (*Dactylorhiza hatigeria*), local tea (*Bidens pilosa*), local caraway (*Carum carvi*), black caraway (*Bunium persicum*), oregano (*Origanum vulgare*), local mint (*Mentha longifolia*), yarrow (*Achillea* sp.) and rose root (*Rhodiola* sp) etc. to prepare herbal tea. The tea is rich in flavonoides, vitamins and has medicinal values (Dwivedi *et al.*, 2006) and has found wide acceptance among the consumer. Several other products such as antioxidant herbal supplement, seapricot beverage, jam, jelly, seabuckthorn oil soft gel capsule, UV protective oil, bakery products, animal feed etc. are at different stages of development and commercialization. The market for seabuckthorn is expanding and industrialists have shown keen interest in the shrub. Seabuckthorn fruit worth Rs 1.4 crore has been sold in 2007 from Leh district of Ladakh region, which generated employment and rural income (Stobdan *et al.*, 2011). This account for less than 5 percent of the region total potential. In year 2007, average collection per household as estimated by us is about 152 kg per year resulting in net annual income of Rs 3344 per household in Nubra block and 117 kg per household per year resulting in net annual income of Rs 2574 per household in Leh block (Stobdan *et al.*, 2011). The demand for fruit is increasing as reflected in the price of its berry from Rs 8/kg in year 2001 to Rs 22/kg in 2010. The shrub is now considered a potential horticultural crop for the region.

9. Seabuckthorn Production and Harvesting

9.1 Soil

Seabuckthorn is found growing profusely on a wide range of soil types. However, soil with light physical structure, rich in nutrient compounds, with a pH near neutral (pH 6.5-7.5) is found to be better. Best growth occurs in deep, well drained, sandy loam with ample organic matter.

9.2 Irrigation

Seabuckthorn can tolerate drought condition but it is a moisture sensitive plant especially in the spring at flowering and fruit development stage. For economic reasons, seabuckthorn orchard establishment should be restricted to areas receiving a minimum of 400 mm annual precipitation, unless irrigated. The optimum soil moisture content for mature plant is around 70 percent. Mulching should be practised to reduce loss of soil moisture (Li and McLoughlin, 1997). The use of black plastic as a mulch during first 3-4 years from seed has been found ideal for moisture conservation and weed control.

9.3 Manure and Fertilizer

Seabuckthorn has a high requirement of nutrient. The plants require little nitrogen and the requirement of potassium is negligible. The plant responds well to phosphorus fertilizer. Well rotten cattle manures @ 45 tonnes/ha should be used depending upon soil condition (Li and McLoughlin, 1997).

9.4 Propagation

Seabuckthorn can be propagated by various methods. The seeds remain viable for more than three years. Freshly harvested seeds have a short physiological dormancy and thus do not germinate immediately after harvesting. Seed stratification for 20-25 days in alternate layers of moist sand yield over 85 percent germination. In Ladakh condition, direct sowing of seeds in November also results in high germination percentage. Cold water soaking of seeds for 6 days yields 93.33 percent germination and 73.33 percent survival rate. Hard wood cutting is a method of choice for large scale propagation. One or two year old pencil thickness cuttings dipped in 2/3 to 1/2 of lower portion in water for 2-3 days followed by 500 ppm IBA treatment results in 85 percent success in Ladakh condition. Planting of the cuttings in greenhouse/trench/polyhouse results in even higher success rate in cold desert condition. Micropropagation of seabuckthorn has been standardized in Murashige & Skoog media. For shoot proliferation 0.01 ppm IBA is found optimum while 2.0 ppm BA and 1.0 ppm NAA is optimum for rooting. Defence Institute of High Altitude Research has successfully standardized asexual method of propagation of seabuckthorn through soft wood cuttings using 200-300 ppm NAA. The technique has yielded over 90 percent rootability in cuttings with excellent root quality under moist condition having course, sand rooting media. The plantlets get ready for transplantation in poly bags within 90 -100 days after planting of cuttings (Dwivedi *et al.*, 2007). Seabuckthorn has profuse suckering habit. A single plant produces 13-65 suckers which can be separated from the mother plant (Dwivedi *et al.* 2006). The individual suckers have been used as true to type planting material.

9.5 Nursery Management

Light and well drained soils are ideal. Soil should be at least 1.2 m deep. The site should receive sun light and irrigation should be assured. Nursery bed should be ploughed deep before sowing. Well rotten Farm Yard Manure (FYM) @ 4kg/m^2 should be incorporated in the soil 20-30 days before sowing/planting.

9.6 Orchard Establishment

Single or double hedge row system is recommended for seabuckthorn plantation under orchard system. In single hedge row system the spacing between row to row is maintained at 2-4 m for pure cultivation and at 4-5 m for intercropping (Dwivedi *et al.*, 2006).

The plants are planted in row at a distance of 1-2 m. Method and spacing need to be modified based on cultivar, soil fertility status and climatic condition. For economic reasons, the ratio of male to female plant is important, as the number of female trees directly affects the total yield. Recommended male plants in an orchard vary from 6 to 12 percent.

9.7 Cultivar

There is no recommended cultivar so far, for cold desert Indian condition. Production characteristics such as yield, ability to harvest, organoleptic test of fruit, nutrient profile of fruit and disease resistance need to be taken into consideration.

9.8 Harvesting

Harvesting of fruit starts soon after ripening. Fruit harvesting is the most time consuming operation in growing seabuckthorn. The relative small fruit size, short pedicel, force required to pull off each fruit, density of fruit on the branch and the thorniness of the plant are the major factors affecting harvesting. Harvesting of fruit is comparatively easier in early morning hours. Fruit can be harvested by various methods. Presently fruits are being harvested by beating the bush. However, mechanical harvester and use of bioregulators need to be encouraged.

9.9 Postharvest and Storage

Seabuckthorn fruits are soft, delicate, juicy and small. Fruits are highly perishable and cannot be transported over long distance. After harvest the berries should be spread in shaded area. Precooling is advisable if the temperature during harvesting is high. Fruits need to be processed after cleaning as soon as possible or can be quick freeze at -18°C for long term storage.

10. State Forest Department Initiatives on Seabuckthorn Development

10.1 Jammu and Kashmir

In Jammu and Kashmir, *Hippophae sp.* is found mainly in Ladakh region ie. Leh and kargil District. Two species viz., *H. rhamnoides* and *H. tibetana* are found in the districts but the majority of the area is under *H. rhamnoides*. In Leh district, the seabuckthorn is naturally found in Nubra and Leh area and also area under seabuckthron was extended by Forest Department since 1960's through root suckers. It is found naturally along Zanskar and Suru Valley of Kargil. It is planted by the Forest Department essentially for soil conservation purposes. Forest department used seabuckthorn in fixing sand dunes in Hunder, Nubra for that more than 70 ha area were planted with seabuckthorn. A total of around 12000 ha is under seabuckthorn (natural as well as planted) in Ladakh region (satellite imagery based finding). Presently it is planted in other ranges/blocks like Durbuk and Nyoma and the results are satisfactory. Presently Forest Department allows people to collect the fruits during the season by giving permits from forest and community area. It allots area on rotational basis. Fruit collection is allowed only upto 10 am in the morning as continued collection during the day after 10 am leads to damage to plants (increased temperature leads to loss of turgidity). The present methodology employed is collection of fruits by beating. There is a detailed regulation on seabuckthorn collection in place in Leh. Co-opertive Dept., Leh collects the fruits from farmers/collectors and pays Rs. 22/kg. Co-operative department makes pulp from fruits and sells it to the industries. As per regulation Rs. 1000/tone of pulp is collected and is kept for seabuckthorn development.

Forest department is acting as a nodal agency for the Ladakh Autonomous Hill Development Council (LAHDC) for the programme KVIC-SFURTI (Scheme of Fund for Regeneration of Traditional Industries) Seabuckthorn cluster development programme. Total cost of the project is ₹ 88.5 lakhs in that Implementing Agency contribution is ₹ 8.5 lakh. Financial assistance is available for i) The creation of Common Facility Centre ii)Product Design and Development iii) Market Promotional Assistance iv) Capacity Building Measures etc. In the cluster development programme 500 families and above are enlisted as beneficiaries. It is planned to develop Common Facility Centre for pulping, product design and packaging. In that fruit pulping machine, bottle sealing machine, labeler and other instruments required will be purchased. Training to farmers, uneducated youths in making seabuckthron juice, jam, sepricot juice, seabuckthron leaf tea and good management Practices has already been started. In the first batch, people from various parts of Leh district were trained in the year 2009 with the help of Defence Institute of High Altitude Research (DIHAR). The major aim of the programme is imparting entrepreneurial skill.

10.2 Uttarakhand

The different species of Seabuckthorn present in Uttarakhand are *Hippophae salicifolia* and *H.tibetana*; *Hippophae salicifolia* is found in large populations in high altitude areas above 2000m, especially along river valleys of rivers or its tributaries originating from glaciers. The major patches identified are Tons valley above Naitwar, Yamunotri valley above Ranachatti, Alaknanda valley above Hanumanchatti, Bhagirathi above Harsil, Nanadakini above Ghat, Pindar valley above Khati, Saryu above Munnar, Gori ganga above Lilam, Dhauliganga West above Suraithota, Dhauliganga East above Sobala and Kali valley above Malpa. On the other hand, *Hippophae tibetana* is found in very rare populations, the major ones being a 100ha patch on Burfu glacier and a 20ha patch on Milam glacier. Small populations are also identified at Bedang in Darma valley 19.92 sq.km of Seabuck thorn forests are in Reserved Forests as per Uttarakhand Forest Statistics 2009-10. The major portion is in Civil lands. A small proportion is also found in private lands.

The species is propagated by seeds as well as by cuttings in many of the high altitude nurseries in Uttarakhand by the Forest Department. 1 sq.km is at present planted with seabuckthorn in forest and non forest areas. Only *H. Salicifolia* is planted.

10.3 Himachal Pradesh

Wild populations of 3 species of Seabuckthorn i.e. *Hippophae rhamnoides* subsp. *turkestanica, H. salicifolia,* and *H. tibetana* have been recorded from Lahaul & Spiti and Kinnaur districts of Himachal Pradesh, the natural zone of occurrence of Seabuckthorn in the State. As per preliminary assessment, the area of occupancy of Seabuckthorn in the State is about 1000 hectares, with geographical spread extending over about 1000 km^2 in the districts of Kinnaur and Lahaul & Spiti. More than 80 percent of the area under Seabuckthorn is designated forest with the remaining 20 percent area being private/community lands.

Seabuckthorn for departmental plantations is being currently raised in 8 nurseries in the State with annual production of about 1.00 lakhs plants. It is pertinent to mention that the usual technique being followed in the State is to raise Seabuckthorn nurseries from stem cuttings. Cuttings maintained in the nurseries for 3 years gave the best survival on planting out. Over the past three years, Seabuckthorn plantations have been raised over 105 hectares in Lahaul & Spiti district.

10.4 Sikkim

In case of Sikkim state, seabuckthorn is mostly confined to Lachen and Lachung valleys of north district and a very few patches are found in East District. The approximate forest area covered under seabuckthorn is about 800 ha. The species found in Sikkim is mostly *H. salicifolia* and it is distributed from 8000 to 12000 ft above msl. The Forest department of Sikkim has planted 220 ha with seabuckthorn. The department has initiated harvesting of berries from 2009-10 and about 4000 kgs of raw juice of seabuckthorn is collected by SHGs during that year.

11. Prospects

Mega projects on seabuckthorn have been initiated by Ministry of Environment & Forests and several R&D organizations in view of its environmental, biotechnological, neutraceuticals, pharmaceutical and socio-economic potential. Traditional usage coupled with commercial value and modern scientific research brings immense benefit to modern society from the lesser known shrub of the Himalayas (Stobdan *et al.*, 2008a).

References

Dwivedi, S.K., Singh, R. and Ahmed, Z. (2005) Morpho-biochemical characteristics of seabuckthorn (*Hippophae rhamnoides*) growing in cold arid Ladakh (India). In *Seabuckthorn Multipurpose Wonder Plant*. 2005, Vol. II. *Biochemistry and Pharmacology.*, New Delhi.

Dwivedi, S.K., Singh, R. and Ahmed, Z. (2006,). *The Seabuckthorn*. Field Research Laboratory, DRDO, Leh, Ladakh (J&K), India.

Dwivedi S.K., Hemraj and Ahmed, Z. (2007). Standardisation of a fast method of vegetative propagation of Seabuckthorn through soft wood cutting in India. In *3rd International Seabuckthorn Association Conference*, Quebec, Canada, August,12-16, 2007. Paper No83.

Gurmet, P. (2009). Seabuckthorn in SOWA-RIGPA (Tibetan medicine), In *Seabuckthorn Hippophae spp.:The Golden Bush*, (eds. Dwivedi, S.K., Parimelazhagan, T., Singh, S.B., Ahmed, Z.), SSPH, Delhi, India, pp. 105-112.

Jike, Z. and Xiaoming Z. (1992) Progress of study on *Frankia* in nodules of Seabuckthorn. *Hippophae*, 2:4-10.

Kala, R., Hema, P.S., Dwivedi, S., Chaurasia, O.P., Banerji, A. 2003. Phytochemical studies on Indian seabuckthorn (*Hippophae rhamnoides*). In *Souvenir of Congress of International Seabuckthorn Association*, p.23, September 14-18, 2003, Berlin.

Li, T.S.C. (2002). Product development of Seabuckthorn, In *Trends in New Crop and New Uses*. (eds. Janick J., Whipkey, A.) ASHS Press, Alexandria, USA.

Li, T.S.C. and McLoughlin, C. (1997). *Seabuckthorn Production Guide.* Canada Seabuckthorn Enterprises Ltd, Canada.

Lu, R. (1992). Seabuckthorn: A multipurpose plant species for fragile mountains. *ICIMOD Occasional Paper No. 20.* Kathmandu, Nepal.

Mathew, D., Parimelazhagan, T., Gomez, S. and Ahmed, Z. (2007) Characterization of seabuckthorn (Hippophae spp.) genetic resources in India using morphological descriptors. *Plant Genetic Resources Newsletter,* 149: 22-26.

Roy, P.S., Porwal, M.C., Sharma, L. (2001). Mapping of *Hippophae rhamnoides* Linn in the adjoining areas of Kata in Labaul and Spiti using remote sensing and GIS. *Current Science,* 80(9): 1107-1111.

Sankhyan, H.P. and Sehgal, R.N. (2005) Morphological variations in the seabuckthorn populations at nine places of Himachal Pradesh. *Bionotes,* 7 (2): 67-68.

Singh, R., Mishra, S.N., Dwivedi, S.K. and Ahmed, Z. (2006) Genetic variation in seabuckthorn (*Hippophae rhamnoides* L.) population of cold arid Ladakh (India) using RAPD markers. *Current Science,* 91(10): 1321-1322.

Singh, V., and Singh, R.K. (2001) Evaluation studies on the Indian and exotic forms of seabuckthorn (*Hippophae* L.). In *Seabuckthorn- A Resource for Health and Environment in Twenty First Century* (V. Singh and P.K. Khosla, Eds.). *Proceeding of International Workshop on Seabuckthorn,* p.107-111, February 18-21, 2001, New Delhi, 274 p.

Singh,V., Gupta, R.K., Uppal, R., Singh, B. and Lata, S. (2005) Morphological and biochemical variations in the seeds of different biotyes of seabuckthorn (*Hippopahe* L.) growing in dry temperate Himalayas. *Non-timber Forest products* (In press).

Stobdan, T., Angchok D. and Singh, S.B. (2008 a) Seabuckthorn: An emerging storehouse for researchers in India. *Current Science,* 94 (10): 1236-1237.

Stobdan T., Chaurasia O.P., Yadav A. and Singh S.B. (2008 b) Seabuckthorn (*Hippophae rhamnoides* L.): A wild shrub of defence and societal importance in cold arid high altitude. *In*: In Service of the Soldier: Life Sciences Perspective. Edited by Manas K. Mandal, K. Ramachandran, A. Gupta. Macmillan Publishers India Ltd.; 388-401.

Stobdan, T. and Singh, S.B. (2009). Gold mine of the cold desert. *Science Reporter,* 46: 39-41.

Stobdan, T., Korekar, G., Dwivedi, S.K., Yadav, A., Targais, K. and Singh S.B. (2009). Research on seabuckthorn (*Hippophae rhamnoides* L.) in India. In *Advances in Agriculture, Environment & Health.* (eds. S.B. Singh, O.P. Chaurasia, A. Yadav, A.M. Rimando, T.H. Terrill). SSPH Delhi, India. pp. 83-97.

Stobdan T, Chaurasia OP, Korekar G, Mundra S, Ali Z, Yadav A. and Singh S.B. (2010). Attributes of seabuckthorn (*Hippophae rhamnoides* L.) to meet nutritional requirements in high altitude. *Defence Science Journal*, 60 (2): 226-230.

Stobdan, T., Chaurasia, O.P., Korekar, G., Yadav, A. and Singh, S.B. (2011). Scope of Seabuckthorn (*Hippophae rhamnoides* L.) for Sustainable Livelihood in Cold Desert of Ladakh, India. In *Global Change, Biodiversity and Livelihoods in Cold Desert Region of Asia*. (eds. K.G. Saxena, L. Liang and X. Xue). BSMS, Dehradun, India. pp. 255-258.

□□□

Chapter-6

Apricot (*Prunus armeniaca* L.) for Livelihood Security in Cold Deserts

Ashish Yadav[1], Tsering Stobdan[1], Sanjay K. Dwivedi[2], Girish Korekar[1], Konchok Targais[1] and R. B. Srivastava[1]

[1]Defence Institute of High Altitude Research (DIHAR), DRDO, Leh, Ladakh

[2]CEPTAM, DRDO, Metcalfe House, Delhi-54

1. Introduction

Agriculture has been in crisis for many years and the world is now facing high inflation and rising food prices. According to Joachim von Braun, head of the International Food Policy Research Institute (IFPRI), "World agriculture has entered a new, unsustainable and politically risky period". The prices of wheat have risen by 77 percent and rice has jumped by 141 percent in 2007 (The Economist, 2008). According to Narayanan (2008), India's food consumption is growing at an annual rate of 1.9 percent, whereas, food production is growing only at 1.2 percent annually. The unequal distribution of food and the world's dwindling natural resources presents a major political and social challenges, likely to reach crisis status with the advancement of climate change and expanson of world population from 6.7 billion to

Keywords: Apricot, *Prunus armeniaca*, cold desert, livelihood security.

9.2 billion by 2050 (Vidal, 2008). The United Nations Educational, Scientific and Cultural Organisation (UNESCO) reports mentioned that the local production of food should be increased using methods that are natural, sustainable and have a minimal impact on the environment. Livelihood security is adequate for sustainable access to and control over resources, both material and social, to enable households to achieve their rights without undermining the natural resource base. The livelihood security approach aims for holistic analysis and understanding of the root causes of food insecurity. Livelihoods are made up of a range of on-farm and off-farm activities, which together provide income and resources to meet basic needs. A livelihood is sustainable when it can cope with sustainable livelihood opportunities now and for the next generation. The high altitude regions (>2500 above MSL) are considered to be one of the most food insecure areas of the world because of their topography, poor soil texture and structure, and harsh climatic conditions. Therefore, the need of the hour is to develop low cost sustainable technologies using natural resources of the high altitudes for the livelihood security and environmental conservation. The high altitude region of Ladakh (32° to 36° North latitude and 75° to 80° East longitude), India, is characterized by extreme cold arid climatic conditions with -40° Celsius temperatures in winter to +35° Celsius in summers, low annual precipitation (80-300 mm), low RH (20-40%), low atmospheric pressure (493 mm Hg), high wind velocity (80-60 km/hr), long frozen winters (5-7 months), very low partial pressure of oxygen and thin atmosphere resulting in heavy influx of ultra violet and infra red radiations (Dwivedi *et al.*, 2007). During winters of about six months from November to April the fields are frozen and nothing can be grown in open, however, some leafy vegetables can only be grown in traditionally designed greenhouses or polyenches. Apricot (*Prunus armeniaca* L.) has been growing in Natural form in Ladakh for over a thousand years. It is the major fruit crop grown in the region on a commercial scale. According to Ladakh Autonomous Hill Development Council (LAHDC), Statistical Hand Book (2006-07) the total area under apricot cultivation is 726 ha followed by apple (*Malus* spp.) 558.25 ha which is the second major crop of the region. The other fruit crops grown on very limited scale in the region are pear (*Pyrus* sp.), peach (*Prunus persica*), plum (*Prunus domestica*), cherry (*Prunus avium*), mulberry (*Morus alba*), and grapes (*Vitis vinifera*). There are no commercial orchards of apricot found in Ladakh; however, it grows naturally in the form of fruit forest in valleys and lower slopes. Apricot is one of the major sources of livelihood and economy for the cold desert regions.

2. Apricot Diversity in Ladakh

Apricot is having very rich genetic diversity in the high altitudes, which can be exploited in terms of fruit morphology, quality and yield for use as table fruit, dehydrated apricot, kernels and oil. According to Dwivedi *et al.* (2007), the prominent apricot cultivars growing in cold arid regions of Ladakh are Amba, Afghani, Australian, Aam, Adzo, Arghoun, Bongti, Brochuli, Badamchuli, Chulikarpo, Charmagz, Chharap, Daychuli, Germoy, Tago, Ghoun Chuli, Kaisi, Gotsos, Gilgit Sweet, Halman, Jeembo Chuli, Ltasmet, Mamur Chuli, Margulam, Moti Tilli, Mircho, Nilipachesmo, Narmo, Narmo Gusmin, Norma Ritonil, Narma Tsangya, Nari, Pechantilli, Papa Chulli, Koban, Khantey, Rakchey Karpo, Rdongmer, Rogan, Shakanda, Shambair, Sheepa-Chesmo, Ston Chulli, Suka, Tripa, Tokpopa, Tokpopa Khantey, Thul Chulli, Urgain Chulli, Yarken, Zan Chulli, Zgoaqchulli etc. Attributed to this high genetic variability, apricot is growing widely in valleys and rugged terrains at different altitudes in Ladakh. Physico-chemical characteristics of some common apricot cultivars of Ladakh are given in Table 1.

Table 1 : Physico-chemical characteristics of common apricot cultivars of Ladakh

Cultivars	Length (cm)	Breadth (cm)	Fruit Weight (g)	Pulp Weight (g)	Stone Weight (g)	Pulp-Stone Ratio	TSS (°B)	Titrable Acidity (%)
Afghani	2.63	2.64	13.2	12.0	1.2	10.0	19.6	-
Australian	4.15	3.86	33.2	30.8	2.0	15.4	15.0	0.92
Charmagz	3.44	3.76	24.8	23.0	1.8	12.8	19.0	-
Halman	2.43	2.43	11.6	10.2	1.2	8.5	18.4	0.67
Raktsey Karpo	3.85	3.82	34.6	32.0	2.4	13.3	21.2	0.61
Safeda	3.35	3.17	19.5	18.1	1.4	12.9	18.0	0.64
Shakanda	3.05	2.89	11.5	10.0	1.5	6.6	19.3	0.95
Shakarpara	3.73	3.58	28.0	25.3	2.7	9.3	12.0	0.13
Tokpopa	3.30	3.34	14.5	12.8	1.7	7.5	12.2	1.00

(Dwivedi *et al.* 2007)

3. Nutritional Values of Apricot

Apricot is considered by many to be one of the most delicious and highly nutritive fruit having pleasant flavour. Fresh and dried apricots having intense golden orange colour are rich source of carotenoids especially β-carotene which represents more than 50% of total carotenoid content (Radi *et al.*, 1997; Sass-Kiss *et al.*, 2005). Besides β-carotene, apricot fruit and its products contain smaller amounts of a-carotene, c-carotene, zeaxantin and lutein (Fraser and Bramley, 2004). Carotenoids are antioxidants that help in preventing heart disease, reduce "bad cholesterol" levels, and protect against cancer. Nineteen amino acids have been identified in apricot and important ones are glutamic acid, aspartic acid, histidine, arginine, tyrosine, cystine, methionine, and hydroxy proline (Faust *et al.*, 1998). Consumption of apricot is very important for healthy eyes, vision, skin, growth, reproduction and immune systems. Apricot also contains soluble fibers and pectin. Soluble fibers help to reduce the risk of heart disease by lowering the blood cholesterol levels and also may help in control of blood sugars. Apricot possesses good laxative properties and is considered to be good for the patients suffering from constipation. In traditional Chinese medicine, apricots are considered helpful in regenerating body fluids, detoxifying and quenching thirst. As early as the year 502 AD apricot seeds were used to treat tumors and in the 17th century apricot oil was used in England against tumors and ulcers. Apricot kernels also have healthy properties, including toning the respiratory system and alleviating the cough. Apricot kernel is an important source of dietary protein as well as oil and fibre (Femenia *et al.*, 1995). The kernel is added to bakery products as whole kernel or grounded and also consumed as appetizers.

Dried apricots are also good source of fiber, iron, calcium and potassium. Dried apricots, contain many of the major vitamins and minerals their are known to be especially beneficial to the functions of the human body, including regulating the proper functioning of the brain, giving energy and help to release stress, help in the recovery of the liver, play an important role in keeping the bones and teeth healthy and strong, and prevention of anemia by increasing blood production, besides, playing an important role in the therapy of ulcers. Apricots also prevent the formation of stomach and intestinal ulcers, decrease the formation of kidney stones, provide a regulatory role in the reproduction system and increase sexual potency, protect against cancer, make cardiac muscles stronger and regulate heart functions (Ackurt, 1998).

4. Apricot for Livelihood Security

Because of higher productivity and income per unit area, horticulture is playing a vital role in livelihood security of the rural masses in cold deserts. More fruit production and additional area brought under it are liable to bring vital changes in the structure of rural economy (Dhiman *et al.*, 2009). Apricot is playing vital role in livelihood and economy of the cold desert regions. Almost every part of the apricot is used by the local inhabitants. Ripe apricot is an excellent desert fruit. It is dehydrated with or without seed since ancient time which is still practiced world over. The fruit is also processed for canning, as candy, bar, beverage, nectar, puree *etc.* The climatic conditions of cold arid desert of Ladakh are highly congenial for the production of export quality apricot, owing long day length, high light intensity, high concentration of UV rays during fruit bearing season, and clear sky, high day temperature followed by low night temperatures and low atmospheric humidity at the time of active fruit growth and maturity period during July and August helps in quality apricot production in this region which is far superior to the other part of the country.

However, the yield of apricot in Ladakh is erratic subjected to the non availability of commercial orchards and following of poor cultural practices. The major part of apricot produced is consumed locally and sold on an individual basis, mainly by women selling their produce in markets. The supply of fresh apricot produce to the other parts of country is very limited mostly because of non availability of proper transportation facilities. Apricot has been sold in various forms *viz.* fresh, dried, kernels, oils etc. The efforts to organize apricot drying and marketing have been successful on a small scale, but need to be scaled up to meet internal and external demand. The disorganized nature of the sector and adherence to certain inefficient production and processing techniques implies that 40-60 percent of the apricot production is wasted. The apricot production that is processed (sun dried) is prone to spoilage due to incorrect picking methods, unhygienic and non-controlled drying procedures (Sabharwal and Singh, 2005). The fruits are generally sun-dried on the roof of houses or on large stones then sold with low added value on local markets. The major disadvantage of traditional drying is the loss of actual fruit colour and fruit becomes partially black or brown which fetches lower prices in the market. Because of lack of hygiene, those dried fruits have a limited access to other markets.

a. Nari

b. Rackey Karpo

c. Australian

d. Halman

e. Khantey

f. Afghani

Fig. 1 : (A-f) Important cultivars of apricot in Ladakh

In the present global climate changing scenario, livelihood security of the farmers generally depends upon environmental security. Apricot helps in environmental conservation by preventing soil erosion and desertification due to high wind velocities in high altitudes mainly because of its deep root systems and it also improves soil health by increasing soil organic matter after defoliations during winters. Apricot succeeds even in sandy and wasteland conditions and can tolerate water stress and temperatures as low as -35°C in dormant stage. The high altitude adaptations of plants are a combination of structural and physiological reactions, which can be clearly correlated to the environmental complex (Mani, 1990).

Ladakh is mainly lying in rain shadow of Himalayas which denies entry to the monsoon clouds. This region constitutes about 85 percent of the Indian cold arid desert where the annual average rainfall is about 50 mm. The main source of irrigation water is the snowfall on the mountains. The local people of Ladakh have developed a small scale farming system where land is irrigated by a system of channels which funnel water from the melted ice and snow of the mountains. It is very well known that plant water uptake depends on soil condition, physiological and environmental factors especially solar radiation and vapour pressure deficit and from the coupling degree of the plant with the atmosphere. Apricot thrives well in the cold arid desert of Ladakh and can sustain production with the low levels of soil moisture. Apricot is considered a drought-resistant crop and exhibits some xeromorphic characteristics, such as the ability to endure water stress in the dry season and the loss of leaves in winter (Torrecillas *et al.*, 1999). Plants have developed physiological responses as well as ecological strategies to cope with water shortages, either by stress avoidance or stress tolerance (Martin *et al.*, 1997). These responses allow them to survive and even to maintain some growth under adverse conditions.

Young apricot plants exposed to slight to moderate water stress conditions develop avoidance mechanisms based on stomatal closure, accompanied by leaf epinasty, which can be considered as a complementary mechanism for regulating transpiration, and both have been recognized as important adaptive mechanisms to drought. Under more severe water stress conditions partial defoliation occurred and osmotic adjustment was triggered as a tolerance mechanism (Ruiz-Sanchez *et al.*, 2000). Soil water deficit and high irradiance had a similar effect on plant water status, both lowering the leaf water potential and canopy conductance. According to Roberts and Rosier (1993), the response of stomatal conductance to vapor pressure deficit is an important factor in deciding the amount of water used by trees.

The above studies clearly indicates that apricot has developed physiological and ecological strategies to counteract the adverse conditions of cold arid deserts and thereby helping in the environmental conservation by judiciously utilizing the available natural resources. Besides the apricot, forests also help in intercepting the precipitations by increasing the humidity in the cold arid deserts. It also provides shelter and food to the wild life, especially birds. DIHAR has recently taken initiatives to green the barren mountains of Ladakh by taking massive afforestation program to plant apricot and apple along with willow and poplar trees at South Pullu, North Pullu etc. to protect the biodiversity and conserve the environment. Results of these studies are still awaited, there is need to thoroughly study and identify novel apricot genotypes which can grow at much higher elevations in the cold arid desert of Ladakh and can sustain production with the limited resources.

Keeping above in view, apricot is a very important commodity for socio-economic upliftment of the local people in Ladakh. We have to organize farmers, the system of cultivation, collection, processing, value addition, packaging, good transportation facilities and marketing with an integrated approach. The monoculture of few cultivars will create serious environmental problems, including the breakdown of resistance and the degradation of soil fertility. Land degradation threatens both sustainable production and food security. Therefore, soil restoration and water conservation strategies are of utmost importance in Ladakh. The challenge now is to add the ecological dimension to the improvement of apricot crop productivity. There is an urgent need for an international research network which can facilitate knowledge and technology sharing in the area of improving integrated apricot farming systems and enhancement of productivity on an environmentally sustainable basis. Defence Institute of High Altitude Research (DIHAR), Leh has made pioneering efforts to develop the modern scientific methods of apricot propagation, orchard management practices and has also identified and conserved the valuable apricot germplasm of the region. Solar driers viz. polyhouse, tent and tunnel driers have been developed for hygienic dehydration of surplus apricot fruits in the region. Methods for value addition in apricot *viz.* osmotic dehydration, intermediate moisture, apricot bar, hurdle technology, squash, nectar, jam, etc., have also been developed to boost the economy of the region. DIHAR is also engaged in the development of modern cultural practices for sustainable apricot production, introduction of high yielding apricot cultivars/varieties and supplying of quality planting materials to the farmers to increase the area under apricot cultivation.

5. Conclusion

Apricot being a major fruit crop of the cold desert of Ladakh has immense potential for livelihood security of locals and environmental conservation. It needs better attention that we should develop technologies based on an integrated farming systems approach that can help to increase productivity in perpetuity without associated ecological harm.

References

Ackurt F. (1998) Alternative new apricot products and the role of apricot in the healthy nourishment. The final report of first apricot council, Publication of Malatya Government Malatya; pp 21-23. (In Turkish).

Dhiman K.R., Bhardawaj D.R., Bawa R. and Vikram A. (2009) Producing world class apples in cold deserts for health and economic security - a case study of regional horticultural research sub-station, Tabo in Himachal Pradesh. In: Singh, S.B., Chaurasia, O.P., Yadav, A., Rimando, A.M. and Terrill, T. (Eds). Advances in Agriculture Environment and Health, Satish Serial Publishing House, New Delhi, pp 99-104.

Dwivedi S.K., Kareem A. and Ahmed Z. (2007) Apricot in Ladakh. Published by Field Research Laboratory, DRDO, Leh-Ladakh, pp 47.

Faust M., Suranyi, D. and Nyujto F. (1998). Origin and dissemination of apricot. Hort. Rev. 22: 225-266.

Femenia A., Rosello C., Mulet A. and Canellas J. (1995) Chemical composition of bitter and sweet apricot kernels. J. of Agricultural and Food Chemistry, 43: 356–361.

Fraser P.D. and Bramley P.M. (2004) The biosynthesis and nutritional uses of carotenoids. Progress in Lipid Research, 43: 228–265.

Mani M.S. (1990) Plant life at high altitude. In: Fundamentals of High Altitude Biology, Oxford & IBH Pubishing Co. Pvt. Ltd., pp 19-25.

Martin M., Morgan, J.A., Zerbi G. and Lecain D.R. (1997) Water stress imposition rate affects osmotic adjustment and cell wall properties in winter wheat. Italian J. Agronomy 1: 11-20.

Narayanan K.K. (2008) Balanced approach between food and biofuels is required: CFTRI. Thaindian News. April 27th 2008.

Radi M., Mahrouz M. and Jaouad A. (1997) Phenolic content, browning susceptibility, and carotenoid content of several apricot cultivars at maturity. Horticultural Science, 32: 1087–1091.

Roberts J.M. and Rosier, P.T.W. (1993) Physiological studies in young Eucaliptus stand in southern India and derived estimates of forest transpiration. Agriculture Water Management, 23: 103–118.

Ruiz-Sanchez M.C., Domingo R., Torrecillas A. and Perez-Pastor A. (2000) Water stress preconditioning to improve drought resistance in young apricot plants. Plant Science, 156: 245–251.

Sabharwal A. and Singh A. (2005) Agriculture and agribusiness- Enterprising Ladakh. Published by LAHDC, Druk Pema Karpo Educational Society Drukpa Trust, and SECMOL. pp 1-35.

Sass-Kiss A., Kiss J., Milotay P., Kerek M.M. and Toth-Markus M. (2005). Differences in athocyanin and carotenoid content of fruits and vegetables. Food Research International, 38: 1023–1029.

The Economist (2008) The new face of hunger. April 17th 2008 edition, www.economist.com

Torrecillas A., Galego R., Perez-Pastor A. and Ruiz-Sanchez M.C. (1999) Gas exchange and water relations of young apricot plants under drought conditions. J. Agric. Sci., 132: 445–452.

Vidal J. (2008). Change in farming can feed world – report. The Guardian, April 16th 2008. www.guardian.co.uk

❑❑❑

Chapter-7

Postharvest Management of Fruits and Vegetables in High Altitudes

Ashish Yadav[1], Tsering Stobdan[1], O.P. Chauhan[2], Girish Korekar[1], Konchok Targais[1] and R.B. Srivastava[1]
Defence Institute of High Altitude Research (DIHAR), DRDO, Leh, Ladakh
[1]Defence Food Research Laboratory (DFRL, DRDO), Mysore

1. Introduction

The diverse agro-climatic zones in the country make it possible to grow almost all varieties of fresh fruits and vegetables in India. Among the horticulture crops, fruit crops recorded a two fold increase in area and production in the last decade. India is the second largest producer of fruits (63.5 MT) from 5.78 million hectare ha area and contributes 11 per cent share in global fruit production. Vegetable crops, occupy 7.80 million hectare ha area, with the production of 125.88 MT (Singh, 2009). In the last one and half decade, India's vegetable production has almost doubled and gross vegetable productivity increased by one and half times. However, only 2 percent of fruit and vegetable production goes for processing, while over 25-30 percent is spoiled due to improper handling and storage and the rest is consumed in fresh form. India is, therefore, incurring a precious loss not only in terms of revenue, but also in terms of health.

Fruits and vegetables play a vital role in human diet as fresh food sources of calories, vitamins, dietary fibre and special nutraceuticals.

In Ladakh due to sub-zero temperature during winter months, fruits and vegetable cultivation is possible only from April to October and harvesting is done from July to October. Round the year supply of fresh food to a large number of troops deployed in Ladakh region especially in forward areas has been a major challenge. Government spends substantial amount on transportation of fresh food to the sector from other parts of the country. In spite of heavy expenditure and best efforts, there is a substantial spoilage of food items and loss of texture and nutrients during transportation. Therefore, postharvest management of fruits and vegetables produced during the summer season is a must for their consumption during winter months in Ladakh sector. Defence Institute of High Altitude Research (DIHAR) of Defence Research and Development Organisation (DRDO) has developed several technologies for postharvest management of surplus fruits and vegetables for off-season consumption to enhance the availability of fresh vegetables during peak winter months.

2. Zero Energy Storage of Apples

The high altitude regions are considered to be one of the most food insecure areas of the world because of their topography, poor soil texture and structure, and harsh climatic conditions. Apple is the major fruit crop next to apricot grown in the region mainly in natural form. There are no commercial orchards of apple found in Ladakh; however, it grows naturally in the form of fruit forest in valleys and lower slopes. Apple is having good genetic diversity in Ladakh. Due to good and quality production, apple is playing vital role in the socio-economic development of Ladakh. Long term storage of apple produced during the summers is a must for its consumption during winter months. Defence Institute of High Altitude Research (DIHAR) has developed the energy efficient low cost technology for long term storage of apple for off-season consumption. The apples were dipped in various $CaCl_2$ concentrations (1,2,3, & 5%) for 5 or 10 minutes and packed in perforated butter paper bag, perforated carton boxes and nylon net bags (Fig. 1). Immediately after packing apples were stored in double layered underground storage structures. Nylon net bags were covered by hay to prevent moisture loss. Results revealed that apple were successfully stored for more than six months during peak winter months without any significant loss in nutritional parameters. Apple dipped in 2 percent $CaCl_2$ for 10 minutes, packed in nylon net bags and covered by hay showed best results in term of long-term storage. This low cost technology for long term storage of apple is helpful to minimize the post harvest losses and increase the availability of fruits in Ladakh during winter months.

(a) (b) (c) (d) (e)

Fig. 1 (a-e) : Energy efficiency preservation of apple under sub-zero temperatures in double layered underground storage structures

3. Traditional Method of Vegetable Storage

Vegetables find important place in the Ladakhi menu. Area measuring 310 hectares is under vegetable cultivation producing 75,440 quintals of fresh vegetables annually in Leh district (Anonymous, 2006). The region remains isolated from rest part of the country from November to May every year due to heavy snowfall. Getting fresh vegetables from outside the region is beyond the reach of people due to high air cargo charges of > ₹ 130/kg. Therefore, the important means of getting vegetables during winter months is by storing the locally produced surplus vegetables for off-season consumption in traditional underground storage structures called as vegetable cellars which are described below :

3.1 Pit

A cone-shaped pit locally known as '*Sadong*' is constructed at ground level in a well-drained location. The size of the pit varies from 5 to 6 feet depth, 3 to 4 feet surface diameter and 5 to 6 feet basal diameter. The pit is made in October ending soon after crop harvesting. Vegetables *viz.* potato, carrot, radish and turnip free from cuts, cracks, bruises, or other insect or mechanical injury are placed in the pit and covered with gunny bag. The pit is filled upto 1 foot from ground level and covered with soil. When vegetables are to be consumed within two to three months, 1 foot gap is maintained between the piled vegetable and the ground surface. It is then covered with long wooden sticks and covered with gunny bags followed by 1.5 feet thick soil layer above the ground. When the weather is warm in late autumn or early spring, the above ground soil layer is reduced to 1 foot. Separate pits are made for storage of potato, carrot, radish and turnip. Storing few vegetables together in the same pit is not uncommon in case the total quantity requirement of a household is large. Under these conditions vegetables can be stored in good conditions for upto 5 months. At the end of the winter months, the remaining vegetables are removed from the pit and the same is filled with soil to carry routine agricultural activities. This method therefore uses the same field for storage during winter and vegetable cultivation in summer.

3.2 Cellar

The cellar locally known as *tsothbang* is a rectangular structure with a small outdoor entrance and a ventilator used for storage of cabbage and root crops. Generally its dimensions are m^3 12 × 10 × 6 (L × B × H). The selection of location is based on that it prevents light entrance in the structure during storage period. The structure is made either in the basement of the house or as separate outdoor underground or semi-underground structure. If the structure is made at basement of the house, opening is generally made from the kitchen using staircase. Humidity inside the structure is maintained by sprinkling water every 15 to 20 days interval. Temperature is maintained by opening or closing the ventilator. The ventilator is closed tightly from November to

February when the outdoor temperature is sub-zero. Depending on the outside temperature the ventilator is opened during day time in October and March onwards. For storage in *tsothbang*, cabbage is harvested along with roots and set them side by side with their roots in 6 inch soil layer spreaded on the floor. Root crops are heaped at a corner of the structure and covered with 3 to 4 inch thick soil layer. This method is good for keeping vegetables for 5-6 months.

3.3 Charches

This method involves storage of fresh onion by hanging them from ceiling. The onion crop is harvested when the bulb is matured and leaves are green. For storage in *charches* the bunch of onion is tied together in the form of knot with the help of green leaves. Long sticks strong enough to bear the load of the onion bunches are hanged parallelly on the ceiling of an unheated storehouse using rope hooks. Bunches of onions are carefully hanged on the sticks. Atmost cares are taken to keep the onion untouched after hanging. Repeated hand touching and hard pressing spoils the bulbs. As per requirement, the bulbs are plugged from one end without disturbing the entire lot. The method is effective for storage of onion up to 8 months from October to May.

4. Minimal Thermal Processing and Freeze Preservation of Vegetables

Though Ladakhi people store vegetables in vegetable cellars for long term, but, this crude method of storing vegetables leads to spoilage to a large extent and significant loss of nutrients. The modern cold storage facilities consume lots of energy and their running costs are also very high. Therefore, the need of the hour is to develop low cost sustainable technologies using natural resources of the high altitudes for the socio-economic upliftment and environmental conservations. DIHAR has taken initiatives to design and establish a zero energy based cold storage facility at Changla top at an altitude of 17,586 ft where temperature remains below zero throughout the year for the storage of minimally processed vegetables. This zero energy storage keeps minimally processed vegetables at sub zero temperatures naturally inside the storage structure. This is helpful in augmenting the supply of locally produced vegetables which is available for the troops during peak winter months. DIHAR has developed technology for thermal minimal processing, freeze preservation and zero energy storage of high demand locally produced vegetables.

4.1 Identification of Surplus Vegetables

During summer, Leh market faces glut of vegetables with not much local consumption which causes the significant losses to the farmers. DIHAR has identified five out of surplus, vegetables *viz.* pea, carrot, cauliflower, french bean and capsicum potential for minimal thermal processing and freeze preservation.

4.2 Minimal Processing of Vegetables

'Minimal processing' describes thermal or non-thermal technologies to process food in a manner to guarantee the food safety and preservation as well as to maintain as much as possible the fresh-like characteristics of fruits and vegetables (Manvell, 1997). However, it is well known that processing of vegetables promotes a faster physiological deterioration, biochemical changes and microbial degradation of the product even when only slight processing operations can be used (O' Beirne and Francis, 2003), which may result in degradation of the colour, texture and flavour (Kabir, 1994; Varoquaux & Wiley, 1994). During minimal processing (including peeling, cutting and grading operations), many cells are broken and intracellular products, such as oxidizing enzymes, are released (Laurila and Ahvenainen, 2002), accelerating the decay of the product. Therefore, proper care has been taken at each and every step to minimize the processing loss of vegetables before freeze preservation.

4.3 Pre-process Handling

Vegetables at peak flavour and texture are used for minimal processing. Postharvest delays in handling vegetables are known to produce deterioration in flavour, texture, colour, and nutrients. Therefore, the delays between harvest and minimal processing have been reduced to retain fresh quality of vegetables prior to freezing. The main steps throughout the processing chain of minimal processing vegetables are washing and disinfection. For this reason, guidelines for packing fresh or minimally processed vegetables generally specify a washing or sanitizing step to remove dirt, pesticide residues, and microorganisms responsible for quality loss and decay (Sapers, 2003). washing of vegetables with running water three times after sorting and removal of the damaged and diseased ones because the quality of minimally processed frozen vegetables depend on the quality of fresh products, since freezing does not improve product quality. Shelling of peas and cutting of carrot, cauliflower, french bean and capsicum in small pieces of approximately one inch size are done manually and washed again two times with running water. Pre-process handling, from the time vegetables are picked until ready to eat, is one of the major concerns in quality retention.

4.4 Blanching

Blanching is the exposure of the vegetables to boiling water or steam for a brief period of time to inactivate enzymes. If vegetables are not heated sufficiently, the enzymes will continue to be active during frozen storage and may cause the vegetables to toughen or develop off-flavours and colours. Blanching of pea, carrot and french beans is carried out with 2 percent salt solution at 100°C for 3 minutes and for capsicum 1.5 minutes. Cauliflower is blanched with 2 percent salt and 0.5 percent sugar solution at 100°C for 3 minutes. Shaking of the blanching basket is done from time to time to distribute the heat evenly. Blanched vegetables promptly cooled down by dipping blanching basket in to large amount of cold water to control and minimize the degradation of soluble and heat-labile nutrients. The enzymes used as indicators of effectiveness of the blanching treatment are peroxidase, catalase and lipoxygenase. Peroxidase inactivation is commonly used in vegetable processing, since peroxidase is easily detected and is the most heat stable of these enzymes (William *et al.*, 1981). A high correlation has been established between development of off-flavours during freezing storage and remaining peroxidase and lipoxygenase activities (Romero *et al.*, 2004).

4.5 Drying

Drying of blanched pea, carrot and cauliflower has been done in thermostatically controlled hot air electric dehydrator at 60°C for 2 hours, for french beans 40 minutes and for capsicum 12 minutes until vegetables were dry enough to prevent microbial growth, subsequent spoilage and it also preserves the flavour, texture and colour of the vegetables (Fig. 2). Water content of dried vegetables varies from 15 to 25 percent depending on the position of trays in drier. The 60°C temperature for 2-3 hours was found optimum for drying as it removes moisture quickly and does not seriously affect the flavour, texture and colour of the vegetables.

Fig. 2 : Drying of Peas

4.6 Packaging

After hot air drying pea, carrot, cauliflower, french beans and capsicum are immediately packed in to 300 gauge polyethylene bags of 500 g size. Proper squeezing of the bags has been done to remove the air as much as possible from the bags before sealing because the excess air leads to flavour loss or off-flavour in frozen foods. Proper packaging is important to protect the product from contamination and damage as well as to preserve food value, flavour, colour and texture.

4.7 Freeze Preservation of Vegetables

Freezing has been successfully employed for the long-term preservation of many foods, providing a significantly extended shelf life. The process involves lowering the product temperature generally to -18°C or below (Fennema *et al.*, 1973). The physical state of food material is changed when energy is removed by cooling below freezing temperature. The extreme cold simply retards the growth of microorganisms and slows down the chemical changes that affect quality or cause food to spoil (George, 1993).

Competing with new technologies of minimal processing of foods, freezing is the most satisfactory method for preserving quality during long storage periods (Arthey, 1993). Most vegetables retain their natural colour, flavour and texture better when frozen than if other methods of food preservation are used. When compared in terms of energy use, cost and product quality, freezing requires the shortest processing time. Any other conventional method of preservation focused on vegetables, including dehydration and canning, requires less energy when compared with energy consumption in the freezing process and storage. However, when the overall cost is estimated, freezing costs can be kept as low as any other method of food preservation (Harris and Kramer, 1975).

The freezing process is a combination of the beneficial effects of low temperatures at which microorganisms cannot grow, chemical reactions are reduced, and cellular metabolic reactions are delayed. The preservation of vegetables by freezing is one of the most important methods for retaining high quality in agricultural products over long-term storage. In particular, the freshness qualities of raw vegetables can be retained for long periods, extending well beyond the normal season.

In our study, air blast freezer was used to freeze the vegetables. The air blast freezer is one of the oldest and commonly used freezing equipment due to its temperature stability and versatility for several product types. Freezing was accomplished by placing the pea, carrot, cauliflower, french bean and capsicum in freezing room. Air blast freezing was done at -18°C for 3 hours until ice crystal formation. Forced air freezing is the cheapest way of freezing which consists of air circulation by convection inside the freezing room and has the added advantage of a constant temperature during frozen storage. However, it is the slowest method of freezing due to the low surface heat transfer coefficient of circulating air inside the room. Freezing time in freezers is largely dependent on the temperature of the freezing chamber, initial temperature and size of product.

4.8 Storage of Freeze Preserved Vegetables

Immediately after freezing pea, carrot, cauliflower, french bean and capsicum bags are stored in -18°C cold storage. Later on these frozen vegetables are shifted to zero energy based cold storage facility created at Changla top (17,586 ft above MSL) for long term storage and off-season consumption during peak winter months (Fig. 3). This permafrost storage keeps the minimally processed vegetables at -15°C to -20°C temperature naturally inside the structure with 10-30 percent moisture during winter months. Studies regarding the shelf life of frozen vegetables stored in zero energy cold storage at Changla top as well as in traditional -18°C cold storage at DIHAR HQ showed no changes in colour, texture, flavour and sensory attributes up to five months of storage. Nutritional profiling and microbial load analysis of minimally processed frozen vegetables revealed that there was no significant loss of nutrients during minimal processing and no microbial growth was observed after even after nine months of storage in -18°C cold storage at DIHAR HQ. However, vegetables stored at Changla top were in good conditions up to April and it started deteriorating from May onwards when temperature rises up to -6°C. Based on results, it can be concluded that minimally processed and freeze preserved vegetables can be successfully stored for five months from December to April in zero energy based cold storage at Changla top without significant loss of nutrients.

Fig. 3 : Minimally Processed Vegetable Storage at Changla Top

5. Development of Value Added Products

The demand for horticultural produce in India is on the rise, owing to increasing populations, changing food habits, the nutritional value of horticultural crops and a greater emphasis on postharvest management, processing and value addition. The unnecessary waste of valuable fruits and vegetables are checked by processing into value added products. Besides, zero energy storage and minimal processing of fruits and vegetables, DIHAR has also developed several value added products from fruits.

5.1 Seapricot Blended Beverage

To counteract the oxidative stress and enhance the performance of soldiers by improving nutrition, a cloud stable blended herbal beverage 'Seapricot' based on pulp of seabuckthorn and apricot has been developed (Fig. 4). The 'Seapricot' is very rich in antioxidants, unsaturated fatty acids, carbohydrate, proteins, minerals and Vitamins A, B_1, B_2, B_3, B_6, B_9, B_{12}, 'C' and 'E'. The user's trial of 'Seapricot' is underway and patent application has also been filed. 'Seapricot' contains 20 : 80 fruit pulp of seabuckthorn and apricot, which forms a sediment-free beverage.

Fig. 4 : Seapricot

5.2 APCOT Herbal Beverage

Apple and apricot are the main fruit crops of Ladakh. Apricot is very rich in carotenoids that help in prevent heart disease, reduce "bad cholesterol" levels, and protect against cancer. Nineteen amino acids have been identified in apricot and important ones are glutamic acid, aspartic acid, histidine, arginine, tyrosine, cystine, methionine, and hydroxy proline. Local apple is also very nutritious and juicy and a good saying is that "An apple a day, keeps doctor away". However, the shelf life of apricot and local apple is very less and without proper transportation and cold chain system most of the produce goes waste and farmer bear's heavy monetary losses. Keeping in view, a cloud stable herbal beverage APCOT has been developed which contains fruit pulp of apple and apricot (Fig. 5). In house studies reveal its shelf life is one year without any sedimentation. The product leads to value-addition to apple and apricot which otherwise get wasted in Ladakh.

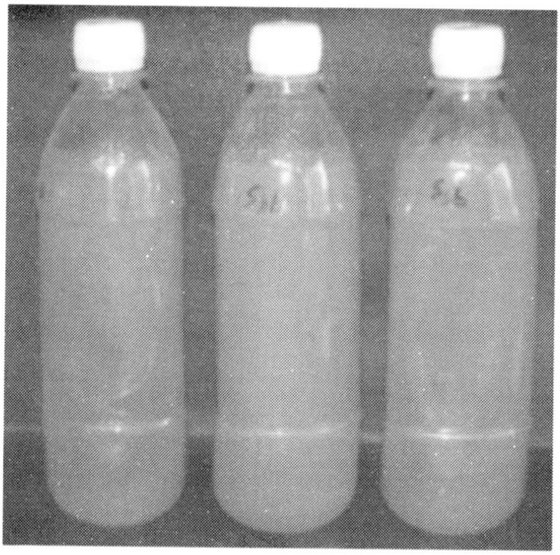

Fig. 5 : APCOT

5.3 Apricot Nectar

Apricot nectar (Fig. 6) is value added product prepared from low quality apricot. It is rich in vitamin A and minerals like calcium (Ca) and Iron (Fe) and has tremendous potential for commercialization. Besides above said products, Apricot Jam, Apricot squash, Seabuckthorn beverages and teas were also developed for enhancing the nutritional security at Ladakh (Fig. 7 & 8).

Fig. 6 : Apricot Nectar

Fig. 7 : Value Added Products from Apricot

Fig. 8 : Value Added Products from Seabuckthorn

6. Conclusion

The fruit and vegetable requirement of Army is huge in Leh-Ladakh and during winter months most of the fruits and vegetables comes by Indian Air Force air craft which involve high cost of transportation. The zero energy storage of apples, minimal processing, freeze preservation and zero energy storage of surplus vegetables and development of value added products will be helpful in boosting the supply of local fruits and vegetables in winters to the troops. All these activities heighten the fresh food availability in Ladakh sector and further strengthen DIHAR's technologies for the benefit of Army and Civil population in Ladakh.

References

Anonymous. (2006). *Statistical Hand Book for the Year 2005-06* (District Statistical & Evaluation Agency, Ladakh Autonomous Hill Development Council, Leh, Government of Jammu & Kashmir, India).

Arthey, D. (1993). Freezing of vegetables and fruits. In: Mallett, C.P. ed., *Frozen Food Technology*, Chapman and Hall, London, UK.

Fennema, O.R. (1973). Low temperature preservation of foods and living matter. In Fennema, O. R., Powrie, W. D., Marth, E. H., Eds, Plenum, New York, Ch. 3.

George, R.M. (1993). Freezing process used in food industry. *Trends in Food Science and Technology*. 4, pp. 134.

Harris, R.S. and Kramer, E. (1975). *Nutrition Evaluation of food Processing*, 2nd ed. Avi Publishing Co. Westport, USA.

Kabir, H. (1994). Fresh-cut vegetables. In A.L. Brods, and V.A. Herndon (Eds.), *Modified atmosphere food packaging*. Institute of Packaging Professionals. pp. 155–160.

Laurila, E., and Ahvenainen, R. (2002). Minimal processing of fresh fruits and vegetables. In W. Jongen (Ed.), *Fruit and vegetable processing*. Cambridge, UK/Boca Raton, FL: Woodhead Publishing Limited/CRC Press LLC.

Manvell, C. (1997). Minimal processing of food. *Food Science and Technology Today*, 11: 107-111.

O'Beirne, D., and Francis, G.A. (2003). Reducing pathogen risk in MAP-prepared produce. In R. Ahvenainen (Ed.), *Novel food packaging techniques*. Cambridge, UK/Boca Raton, FL: Woodhead Publishing Limited/CRC Press LLC. pp. 231-232.

Romero, D. M., Castillo, S. and Valero. D. (2004). Quality control in frozen vegetables. In Hui, Y.H. *et al.* ed., *Handbook of vegetable preservation and processing*, Marcel Dekker Inc., New York, USA.

Sapers, G. M. (2003). Washing and sanitizing raw materials for minimally processed fruit and vegetable products. In J. S. Novak,G. M. Sapers, & V. K. Juneja (Eds.), *Microbial safety of minimally processed foods*. Boca Raton, FL: CRC Press LLC. p. 222.

Singh, H.P. (2009). Dynamics of technologies in horticulture for improving production, productivity and profitability. *Proceedings of International Conference on Horticulture, Bangalore*, Karnataka, India Nov. 9-12: 80-98.

Varoquaux, P., and Wiley, R. (1994). Biological and biochemical changes in minimally processed refrigerated fruits and vegetables. In R. C. Wiley (Ed.), *Minimally processed refrigerated fruits and vegetables*. New York: Chapman and Hall. pp. 226–268.

William, D.C, Miang, H.L., Chen, A.O., Pangporn, R.M. and Whitaker. J.R. (1981). *Blanching of vegetables* for freezing-which indicator enzyme to choose. *Food Technology*, 40: 130.

❑❑❑

Chapter-8

Trans-Himalayan Medicinal Plants and Herbal Nutraceutical Products

O.P. Chaurasia, Sunil K. Hota, Raj Kumar, Amol B. Tayade and R.B. Srivastava
Defence Institute of High Altitude Research (DIHAR), DRDO, Leh, Ladakh

1. Introduction

India has one of the oldest, richest and the most diverse cultural traditions that is use of medicinal plants and it represents a striking example of the intimate linkage between biodiversity and cultural diversity. The history of medicinal plants can be traced back to the Vedic period (4500 B.C. to 1500 B.C.). The identity of several plants like Semal, Pithuan and Pipal referred to in the *Suktas* of *Rig Veda* can be fixed with reasonable certainity. The *Atharveda* contains detailed information on approximately 2000 medicinal plants and their uses. After the Vedic era, the works of Charak and Susruta namely *Charak Samhita* and *Susruta Samhita* deal with 700 drugs both of daily and specific uses. The period of the 16th and 17th century witnessed a spurt in the development of medical botany and most of the world famous medical drugs came into light during this period. For this reason, the period is also referred to as "The Age of Herbal Medicines". In India, over 7500 plants species are estimated to be used by approximately 4635 ethnic communities for human and veterinary health care across various ecosystems ranging from Ladakh in the trans Himalayas cold desert to the southern coastal tip of Andaman & Nicobar Islands and from the deserts of Rajasthan & Kutch to the Hills of the North East (Chaurasia *et al.*, 2007; Irshad *et al.*, 1989).

The people of Ladakh and Lahul-Spiti of the trans Himalayan region have their own medical system called "Sowa Rigpa System of medicine" (Amchi system or Tibetan Medicine) and the practitioners are called Amchis (Superior of all). Amchi medicine is based on skillful use of plants, minerals and animal products. The origin of Amchi medicine is traced back to more than 2500 years ago in India, when Lord Buddha delivered a medical practice called Bzi (Chatus tantre) which was translated in Tibetan language during 8th century AD by Acharya Chandrananda and Varochana. In order to popularize these texts in Tibet, the senior Yusthog Youthan Gonbo had made some changes in it to make it harmonious with the culture and environment of Tibet. Since then this system is very popular in Tibet and in most parts of the Himalayan belt. This system was introduced in cold arid zones during 10th -11th century AD.

The Amchi system of medicine has been in vogue in Ladakh and other parts of Trans Himalayas since centuries. Amchis are looking after 60 percent of public health. Amchis not only have theoretical texts but also practical experiences aquired from several generation. They also have a reputation of having high ethical standards in the social system. Generally single herbal preparation and combinations of two or more herbs are prescribed and administered (Sharma, 1989; Mukherjee *et al.*, 2006; Singh and Chaurasia, 2000).

2. Ethnobotanical Studies

The extensive ethno-medico-botanical explorations revealed that Ladakh and Lahaul-Spiti of the Trans Himalayas are abundant in medicinal plants and approximately 35 percent of total flora are of medicinal value. The widely used medicinal plants in traditional systems of Indian medicines including Sowa Rigpa system (Amchi system) are *Aconitum heterophyllum, Dactylorhiza hatagirea, Ephedra gerardiana, Hippophae rhamnoides, Inula racemosa, Origanum vulgare, Rheum webbianum, Rhodiola imbricata, Saussurea lappa etc.* (Anonymus, 1976; Kachroo *et al.*, 1977; Kaul, 1997; Stewart *et al.*, 1976). The ethno-medico-botanical studies conducted in Trans-Himalayan region revealed various categories of medicinal plants which are described below :

Source of Vitamins	:	*Hippophae rhamnoides, Achillea millefolium, Dactylorhiza hatagirea*
Source of alkaloids	:	*Aconitum hetetophyllum, Ephedra gerardiana, Podophyllum hexandrum*
Source of saponins	:	*Dactylorhiza hatagirea, Primula* sp.
Source of mucilage	:	*Malva* sp., *Plantago* sp., *Ulmus* sp.
Source of phenols	:	*Origanum vulgare, Thymus vulgaris*
Source of fixed oils	:	*Hippophae rhamnoides, Prunus armeniaca*
Source of essential oils	:	*Dracocephallum heterophyllum, Perovskia abrotanoides*

The present chapter deals with brief description of the plant species of nutraceutical values, their families, English (E), Hindi (H) & Local names (L), their traditional and other uses and the neutraceutical formulations comprising of these plants.

3. Nutraceutical Plants

3.1. *Achillea millefolium* Linn. (Asteraceae)

Name	:	Gandana (H), Milfoil/Yarrow (E), Chuang (L)
Plant Characters	:	A perennial herb. Stem branched, furrowed reaching up to 50 cm tall, pubescent, glabrous. Leaves narrowly oblong lanceolate. Flower heads terminal, corymbose, white short pedicel.
Uses	:	Traditionally the extract of leaves is useful in urinary problems, toothache and gum inflammation. The whole herb is credited with astringent, stimulant, tonic, diaphoretic, cold, colic, heartburn, hysteria, epilepsy and rheumatism. The herb contains alkaloids and quaternary bases like achilleine, achilletin, betaine, betonicin and choline. It is one of the constituents of the Ayurvedic preparation 'Liv-52'.

3.2. *Allium przewalskianum* Regel. (Amaryllidaceae)

Name	:	Wild garlic (E), Skotche (L)
Plant Characters	:	A small perennial bulbous herb. Bulbs elongated, clustered, scales finely reticulate, red or brown. Leaves several, hollow, succulent fragrant,

		reaching up to 60-100 cm height. Flowers several in heads, pink.
Uses	:	Traditionally leaves are edible; used as condiment and as vegetable. The leaves are made into paste, which is consumed either fresh or dried and made into round balls for winter use. These balls are usually available in local market. The dried leaves are used against indigestion.

3.3. *Arnebia euchroma* (Royle.) John. I. M Johnston. (Boraginaceae)

Name	:	Ratanjot (H); Demok (Dre-mog) (L)
Plant Characters	:	A tufted very bristly haired perennial herb with stout rootstock. Stem many, arising from the rootstock, 25-40 cm. Leaves linear to narrow lanceolate, bristly haired. Flowers many in rounded cluster, pale pink-purple turning blackish-purple.
Uses	:	In Amchi system of medicine the root is used as blood purifier, against blood disease, cold, cough, lung and pulmonary problems. The roots are also used as hair tonic by the locals. The roots contain Ac-Shikonin and four related compounds - alkamin-B, B-di-Me-acrylate. It is reported to be antipyretic, anti-microbial & anti-inflammatory.

3.4. *Artemisia dracunculus* Linn.; *A. glauca* Pallas ex Willd. (Asteraceae)

Name	:	Little Dragon Mugwort (E), Shersing or Burtse (L)
Plant Characters	:	An erect, strong aromatic, perennial herb. Stem grooved, smooth, up to 40-80 cm. Leaves entire, linear-oblong and acute. Flower-heads globular, short-stalked, forming long axillary raceme.
Uses	:	The leaf or flower extract is traditionally used against toothache, urinary problems and sometimes against intestinal problems. Two compounds namely capillene and 3(2-butynyl) isocumarin have been reported to be isolated from roots. It is diuretic, controls menstrual cycle, relieves toothache & intertinal worm infections.

3.5. *Capparis spinosa* Linn. (Capparidaceae)

Name : Ber (H), Caper bush (E), Kabra (L)

Plant Characters : A diffuse and prostrate, perennial shrub or climber with yellow hooked spines. Leaves variable, ovate to elliptic with a minute sharp point at the tip. Flowers large, white to purple. Fruits fleshy, oblong-ellipsoid, with red flesh and many brown seeds.

Uses : Bark and roots are traditionally used against liver problems, rheumatism, tubercular gland disorders and paralysis. The plant extract is used in treatment of viral hepatitis, cirrhosis of liver and old age ailments. It is one of the constituents of the Ayurvedic preparation 'Liv 52'. The root bark contains stachydrine, rutic acid and volatile compounds with odour of garlic. The plant contains glocosinolates, glucocapparin and glucocapangalin.

3.6. *Carum carvi* Linn. var. *gracile* (Apiaceae)

Name : Kala Jeera (E) Umbu, Kosnyot (Go- sNyod) (L)

Plant Characters : An annual or biennial glabrous herb with procumbent or erect stem, 30-60 cm. Leaves bi-pinnately dissected, linear, bracteate with 5-15 white pink flowers. Flowers white or rosy, in dense umbels. Seeds light to dark brown.

Uses : In Amchis system of medicine fruits and seeds are used as febrifuge, to improve eye vision and digestion. It is also useful in hot disorders, and poisoning. The seeds contain a mixture of ketones, carvone, terpene and trace of carvacrol.

3.7. *Dactylorhiza hatagirea* (D. Don.) Soo.; *O. latifolia* Linn. (Orchidaceae)

Name : Salampanja (H), Spotted Heart Orchid (E), Ambolakpa (L)

Plant Characters : A tuberous, perennial herb. Roots palmately divided into 2 to 5 finger like structures. Stem tall upto 70 cm. leaves several, oblong-lanceolate upto 25 cm. Flowers pink or purple in dense flowering spikes.

Uses : Amchis use the tuber roots for health tonic. Tubers are astringent, expectorant, used as nervine tonic, aphrodisiac, useful in diarrhea, bone fracture and chronic fevers. The tuberous roots contain starch, mucilage, sugar, albumen and potassium.

3.8. *Ephedra gerardiana* Wall. ex Stapf. (Ephedraceae)

Name : Asmania/Somlata (H), Ephedra (E), Chhapat (L)

Plant Characters : A low growing tufted shrub. Stem tall 30-60 cm, erect, green, branched. Branches scaly on joints. Male cones ovate, solitary. Female cones usually solitary. Fruits ovoid, red, fleshy bracts. Seeds 2, black.

Uses : The roots, stems and branches are used as a tonic, against fever, stops bleeding, blood purification, snuffing, hepatic diseases, rheumatism and bronchial asthma. The fruit and dried stem constitute a valuable drug "Ephedrine" which is used to cure Asthma. It is a cardiac stimulant and is very effective against asthma, heart failure and rheumatism.

3.9 *Hippophae rhamnoides* Linn. spp. *turkestanica* Rousi. (Elaeagnaceae)

Name : Seabuckthorn (E), Tsermang or Sastalulu (L)

Plant Characters : A dwarf, very thorny, perennial shrub or small tree. Stem woody, erect, long upto 10-15 ft with silvery-waxy covering on the old shoots. Leaves variable, oblong blunt to green above and silvery scaly on both sides. Flowers very small greenish or yellowish. Fruits subglobose, succulent, orange or red.

Uses : The fruits, leaves and seeds have anti-aging, anti-cold, and energizing properties. The fruits are antitusassive, blood purifier, heal peptic ulcer, oral mucocitosis, rectal mucocitosis, lung disorders, cuts and wounds. Used against pulmonary disorders, inflammation of pulmonary tract, expelling of phlegm. Clotting of blood, improves digestion and is useful for spleen disorders.

Seabuckthorn is considered as a storehouse of various vitamins viz. C, B, E, K, Carotenoids which contribute to its strong antioxidant property. The berries are useful in gum bleeding, regeneration of tissues and repair of mucus membrane and other organs. In addition seeds and fruits contain valuable fatty oil which is a rich source of beta-carotene, Vitamin E, rare fatty acids and phytosterols having great demand in pharmaceutical industries. Seabuckthorn has got a valuable importance in Ladakh and other parts of Trans-Himalayan regions after its utilization on commercial scale since 2001. Defence Institute of High Altitude Research developed, patented and commercialized multi-vitamin herbal beverage, Jam and Herbal tea from Seabuckthorn for the first time. The recent products from Seabuckthorn are Herbal anti-oxidant supplement and Seabuckthorn seed oil soft gel capsule (Stobdan *et al.*, 2008 & 2009).

3.10. *Hypericum perforatum* Linn. (Hypericaceae)

Name : St John's Wort (E)

Plant Characters : An erect, branched, perennial herb, upto 60 cm tall. Leaves oblong or ovate-oblong, pellucid-punctate. Flowers yellow in terminal corymbs. Petals black dotted on the margins.

Uses : Traditionally paste of leaves is used in treatment of wounds and cuts. Infusion of fresh flowers in olive oil is used externally for sores, wounds, ulcers, swellings and rheumatism. The herb contains hypericin, pseudohypericin, flavanoids - quercitrin, rutin, isoquercitrin etc. The essential oil contains caryophylene, pinene, limonene, hypericins etc.

3.11. *Inula racemosa* Hk. f. (Asteraceae)

Name : Puskarmool (H), Manu (L)

Plant Characters : An erect, robust, hairy, perennial herb. Leaves leathery, rough above, densely hairy beneath, toothed. Flower-heads yellow, many in racemes, borne on spike like clusters.

Uses : Roots are considered to be useful in bronchial asthma when mixed with Kuth. The roots are used in gastrointestinal disorders and rheumatism. Alantolactone, β sitosterol and its glucosides have been reported in alcoholic extract of roots.

3.12. *Mentha longifolia* Linn. Hudson. (Lamiaceae)

Name : Jangali Pudina (H), Horse Mint (E), Phololing (L)

Plant Characters : An erect, aromatic, perennial herb. Stem hairy, delicate, much branched 30-90 cm long. Leaves mostly basal, sessile, ovate toothed, round to kidney shaped, oblong-lanceolate, 2-7 cm, serrate. Flowers tiny, lilac, borne on terminal spikes.

Uses : The aromatic parts are commonly used as chatani and as flavouring agent to local dishes. It has 0.3-0.5 percent essential oil which contains piperitone oxide, diosphenol, diosphenolene. The plant is considered carminative and stimulant.

3.13. *Origanum vulgare* Linn. (Lamiaceae)

Name : Wild oregano/Marjoram

Plant Characters : A strong aromatic, spreading, perennial herb 30-75 cm high. Stem branched from the base. Leaves broadly ovate. Flowers pinkish or whitish crowded in short dense terminal clusters.

Uses : The plant contains essential oils, tannins, resin, sterols and flavonoids. The major components are thymol, carvacrol etc. Both carvacrol and thymol are antibacterial and antifungal. Traditionally the plant is quite useful as carminative and stimulates the flow of bile. The aromatic plant parts are used as flavouring agent especially for making pizzas. The aerial parts contain essential oil ranging between 0.25-0.40 from dried parts.

3.14. *Podophyllum hexandrum* Royle (Berberidaceae)

Name : Himalayas Mayapple (E), Vankakri or Vanvaigan (H) Demokusu, Papra, Ol-mose (L)

Plant Characters : A perennial scapigerous herb. Stem modified into underground rhizomes. Leaves 2, blade rounded in outline, 10-25 cm, deeply cut into 3 ovate toothed lobes. Flowers solitary, terminal white or pink, cup-shaped. Fruit large, scarlet or reddish berry, pulpy, orange or red seeds.

Uses : Traditionally the plant is used for gynaecological diseases like menstrual irregularity, diseases of the uterus and improves Lung function and blood circulation, helps parturation. The rhizomes and roots of the plant, constitute the drug "podophyllin", Podophyllotoxin, is considered a cholagogue, purgative, emetic and a bitter tonic. Roots have potential chemicals, which are effective in skin cancer and radiation.

3.15. *Rheum webbianum* Royle (Polygonaceae)

Name : Himalayas Rhubarb (E), Revanchini (H), Lachhu or Chu-rtsa (L)

Plant Characters : A stout, erect, perennial herb. Roots thick. Stem veriable, 30-150 cm. Lower leaves long petioled, orbicular-cordate 15-60 cm across. Flowers pale yellow small in cluster. Fruits broadly oblong notched at both ends.

Uses : The roots, stems, leaves and leaf-stalks are purgative and traditionally used in treating indigestion, abdominal diseases, astringent, boils, wounds and flatulence. The roots are an excellent source of anthoquinones, useful as laxative.

3.16. *Rhodiola imbricata* Edgew.; *Sedum roseum* (Linn.) Scop. (Crassulaceae)

Name : Rose root or Stone crop (E), Shrolo (L)

Plant Characters : A succulent, perennial herb 10-35 cm with rose scented massive rootstock. Leaves 1.3-3 cm long, oblong to narrow, elliptic, nearly entire. Flowers pale yellow in congested clusters, surrounded by an involucre of leaves.

Uses : Traditionally the roots are used in medicine against lung problems, cold, cough, fever, loss of energy and pulmonary complaints. Rhodiola plants are excellent source of salidrosides, rosavin, rosarin compounds, having adaptogenic properties.

3.17. *Rubia cordifolia* Linn. (Rubiaceae)

Name : Manjith (H), Indian Madder (E) & bTsod (L)

Plant Characters : A herbaceous climbing perennial. Roots perennial. Leaves smooth or scabrid, whorled, petioled, ovate, acute. Flowers small, white or greenish.

Uses : Traditionally the stems and roots are used against hematologied diseases, chest complaints, discoloration of skin and leucoderma, inflammation, menstrual disorders, parapysis, stomachache, tonic, ulcers, urinary complaints. The roots contain purpurin, munjistinn, pupuroxanthin and anthraquinones.

3.18. *Saussurea lappa* (Decne.) Sch.-Bip.; *S. costus* (Falc.) Lipsch. (Asteraceae)

Name : Costus (E), Kuth (H), Rustha (L)

Plant Characters : An erect, robust perennial herb. Root possessing a characteristic penetrating odour; radical leaves with long lobately winged; flowers dark blue-purple in axillary and terminal clusters; achene curved, compressed.

Uses : The roots are used in bronchitis along with equal quantity of Manu. The roots are being used in asthma, dysentery, rheumatism and ulcer. The root contains an alkaloid Saussurine and tannins.

3.19. *Thymus serpyllum* Linn. (Lamiaceae)

Name : Jangli Ajwain (H), Tumbrak (L), Wild thyme (E)

Plant Characters : A strong aromatic, spreading, tufted perennial herb. Stem branched from the base. Leaves sessile, elliptic-oblong. Flowers pinkish or reddish crowded in short dense terminal clusters giving a look of the field.

Uses : The aromatic plant parts are used as flavouring agent. The aerial parts contain an essential oil known as oil of wild thyme ranging between 0.25-0.40 percent from dried parts. The major components are thymol, carvacrol, etc. The plant in having antispects, antispasmatic and carminative properties.

3.20. *Tribulus terrestris* Linn. (Zygophyllaceae)

Name : Gokhroo (H), Land caltrop or Puncture vine (E), Ze-ma (L)

Plant Characters : An annual, prostrate herb. Stem branched, slender, decumbent, hairy. Leaves pinnate compound, leaflets 4-12 ovate to obovate. Flowers solitary, yellow in long raceme. Calyx ovate to linear. Fruits of 5 stiff achene fused together; each achene triangular.

Uses : Fruits are considered useful in liver aliments, leucoderma, impotency, cold, cough, urinary disorders, and gout. It forms a constituent of the well known Ayurvedic medicine Dashamoola-rishtha and Amritha Prasa Ghritha, prescribed for several diseases. The plant contains carboline alkaloids, steroidal sapogenins, saponins, flavonoids.

4. Herbal Nutraceuticals

Good nutrition and healthy living especially in high altitude regions require a balanced diet. It is the combination of food in our diet that keeps us healthy, and plants provide a large proportion of our daily energy and nutritional needs. The main nutrients are carbohydrates, fats and proteins. These provide all our energy needs. Similarly Vitamins and minerals are present in small amounts but they are very much important for maintaining specific functions of the body. The vitamins of relevance to plant derived foods are divided into two main groups : water soluble vitamins (C, B_1, B_2, B_3, B_5, B_6, B_9, B_{12}) and fat soluble vitamins (A, D, E, K). Similarly important minerals related to growth of our body are calcium, phosphorus, sodium, potassium, chloride, fluoride, iron, copper, selenium, zinc, iodine, magnesium, manganese, etc. The fat is an important energy source in the human diet and is also important for normal growth and development. The essential unsaturated fatty acids viz. Oleic acid (omega-9), Palmitoleic acid (omega-7), Linoleic acid (omega-6), Linolenic acid (omega-3) are essential for the body development. The deficiency of vitamins, minerals and essential fatty acids causes several health diseases viz., heart diseases, cancer, scurvy, megaloblastic anaemia, skin diseases, blindness, deformity in bones, poor wound healing, *etc*. The vitamins like Thiamin, Niacin, Riboflavin and Pantothenic acid play a central role in the production of sufficient energy for thermogenesis. Vitamin C, vitamin B_6, vitamin B_{12}, biotin or folic acid are also useful to combat cold and hypoxic conditions, prevalent at high altitude. Deficiency of a number of trace elements including iron, zinc, or copper may lead to defective thermoregulation (Ben-Erik van Wyk, 2005).

DIHAR has formulated various nutraceutical products based on ethnobotanical information and using Trans-Himalayan medicinal plants (Ballabh *et al.*, 2007; Ballabh *et al.*, 2008). The briefs of nutraceutical products are as under :

4.1 Seabuckthorn Herbal Tea

The anti-oxidant enriched herbal health tea formulated from seabuckthorn leaves (*Hippophae rhamnoides var. turkestanica*) and other high altitude medicinal plants viz. *Achillea millefolium* (Yarrow), *Bunium persicum* (Black caraway), *Carum carvi* (Caraway), *Mentha longifolia* (Wild mint), *Rhodiola imbricata* (Arctic root), *Rosa webbiana* (Wild rose) with minimum processing is rich in flavonoides, vitamins and natural antioxidants. It has medicinal value *viz.* soothing, energizer, anti-stress, stimulant and useful as health tonic. The herbal tea has its own natural colour, flavour and taste. It is presently marketed by M/s Arctic Bio Pro Ltd., New Delhi and M/s Victory Corporate & Technical Services

Ltd., New Delhi. The product is available in market by brand name Si-Berry Tea (Ballabh & Chaurasia, 2007).

4.2 Seabuckthorn Herbal Anti-Oxidant Supplement

Nutraceutical herbal anti-oxidant supplement is formulated using seabuckthorn pulp and extracts of indigenous medicinal plants of high altitude Himalayas. The formulation is rich in Vitamin C, Vitamin E and carotenoids. The technology has been transferred to Indian Medicines Pharmaceutical Corporation Ltd (IMPCL), a PSU of Min. of Health, Govt. of India for commercial production.

4.3 Multivitamin Seabuckthorn Herbal Beverage

The Multivitamins herbal beverage has been formulated using seabuckthorn pulp and is rich in ascorbic acid, tocopherol, Vitamin K Niacin, Riboflavin and Folic acid. Technology of formulation of multivitamin seabuckthorn herbal beverage has been transferred to seven vendors for commercial production. The product is available in market by various brand : Leh Berry, Ladakh Berry, Power Berry.

4.4 Seabuckthorn Soft Gel Capsule

This Institute has formulated Seabuckthorn soft gel capsules using seabuckthorn seed oil extracted through super critical CO_2 technique. The product is entirely herbal and rich in all omega fatty acids along with vitamin E and beta carotene that add to its antioxidant capabilities.

It is beneficial for the heart and blood circulation and increases myelination of neurons, thereby improving cognitive functions. It is rich in anti-oxidants and ameliorates oxidative stress induced biochemical changes thereby resulting in better acclimatization to high altitude.

4.5 Herbal Appetizer

Loss of appetite and indigestion are one of the most common problems in high altitude conditions due to low oxygen pressure, unbalanced nutritional diet, harsh climatic conditions, etc. The herbal appetizer has been formulated from a standardized proportion of different plants parts using *Achillea millefolium* (Yarrow), *Carum carvi var gracile* (Caraway), *Capparis spinosa* (Caper bush), *Foeniculum vulgare, Hippophae rhamnoides* (Seabuckthorn), *Medicago sativa* (Alfalfa),

Flavonoides	Phenols	Carotenoides	Anti-oxidant Value
3.42 %m/m	15.00 mg/g	1.0 %m/m	1200 µg/ml

HERBAL TEA

Vitamin C	Vitamin E	Thymine	Calcium
120 mg/100g	2.3mg/100g	0.21mg/100g	174mg/l

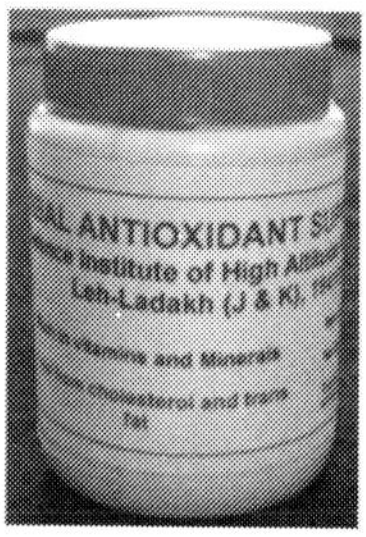

HERBAL ANTIOXIDANT SUPPLEMENT

Aloin	Thymol	Carvone	t-Anethole
0.82%	8.00%	7.00%	7.30%

HERBAL APPETIZER

Oleic acid	Linoleic acid	Linolenic acid	Palmitoleic acid	Carotenoid content
48.66%	35.56%	0.43%	0.53%	84 ppm

UV PROTECTIVE OIL

Oleic acid	Linoleic acid	Linolenic acid	Palmitoleic acid	Carotenoid content
48.66%	35.56%	0.43%	0.53%	84 ppm

SOFT GEL CAPSULES

Fig. 1 : Herbal Neutraceutical Products of DIHAR

Mentha longifolia (Wild mint), *Origanum vulgare* (Origanov), *Prunus armeniaca* (Apricot), *Rheum webbianum* (Rhubarb), *Saussurea lappa* (Costus). The detailed biochemical profiling of herbal appetizer has been conducted and is found rich in aloin, thymol, carvone t-anethole, etc. The Herbal appetizer has carminative, laxative and appetite enhancer properties and found to be very useful in high altitude conditions.

4.6 Herbal UV Protective Oil

Presence of ultra violet radiations (UV) at high altitude causes skin damage and sun burns. To prevent UV mediated skin damage Defence Institute of High Altitude Research has developed a UV protective herbal oil formulation. The herbal UV protective oil is formulated from a standardized proportion of oil extracted from *Hippophae rhamnoides, Prunus armeniaca* and *Seasemae indicum* along with other high altitude medicinal plants containing UV protective metabolites. Herbal UV Protective Oil has topical antioxidant, UV reflective, skin nourishing and moisturizing properties which protects the skin against harmful UV radiations.

4.7 Herbal Adaptogenic Appetizer

Since hypoxia induced hypophagea results in nutritional compromise in high altitude conditions leading to impaired physical & cognitive performance, DIHAR has developed a herbal adaptogenic appetizer comprising of *Prunus armeniaca, Hippophae rhamnoides* and *Rhodiola imbricata.* The formulation not only improves appetite and stabilizes abnormal gastric motility at high altitude, but also augments physical performance and endurance to high altitude conditions.

5. Conclusion

Though the medicinal properties of the high altitude plants is well documented, the sustainable utilization of this natural resource in developing formulations for augmenting acclimatization in high altitude conditions needs to be optimized. Besides that, the mechanism of action of these plants extracts, their long term side effects and the active principle responsible for their therapeutic efficacy need to be determined. Isolation of active principles for therapeutic purposes could result in requirement of lower doses for effective treatment. A concerted effort for conservation, propagation and *in vitro* tissue culture of these endangered plants is also required to ensure sustainable utilization of these natural resources. With the advent of modern biotechnological tools, micro-propagation and cloning of genes responsible for synthesis of the active principles into agricultural plants can be a strategy for economic utilization of the medicinal values of these plants. Besides that scientific validation of the traditional systems of medicine could provide an impetus to herbal products and nurture these age old traditional medicinal systems.

References

Anonymous, (1948-76). *The Wealth of India- Raw Materials.* I-IX. C.S.I.R., New Delhi.

Ballabh, B. and Chaurasia, O.P. (2007). Traditional medicinal plants of Ladakh used in the treatment of cold, cough and fever. *Journal of Ethnopharmacology.* Vol. 112, 341-349.

Ballabh, B. Chaurasia, O.P. and Ahmed, Z. (2007). Herbal products from high altitude plants of Ladakh. *Current Science.* vol. 92. No.12. 1664-1665.

Ballabh, B., Chaurasia, O.P., Ahmed, Z. and Singh, S.B. (2008). Traditional medicinal plants of cold desert Ladakh-Used against kidney and urinary disorders. *Journal of Ethnopharmacology,* 118 No. 2: 331-339.

Ben-Erik van Wyk. (2005). Food Plants of the World. MC Times Editions, Singapore.

Chaurasia, O.P., Ahmed, Z. and Ballabh, B. (2007). *Ethnobotany and Plants of Trans Himalayas.* Satish Serial Publication, New Delhi ISBN : 81-89304-33-X.

Irshad A., Navchoo, A. and Buth, G.M. (1989). Ethnobotany of Ladakh, India. *Journal of Ethnopharmacology,* 26 : 137-146.

Kachroo, P., Sapru, B.L. and Dhar, U. (1977). *Flora of Ladakh.* Bishen Singh Mahendra Pal Singh, 23-A Connaught Place, Dehra Dun.

Kaul, M.K. (1997). *Medicinal Plants of Kashmir & Ladakh (Temperate & Cold Himalayas).* Indus Publishing Company, New Delhi.

Mukherjee, P.K., Wahile and Atul. (2006). Integrated approaches towards drug development from Ayurveda and other Indian system of medicines. *Journal of Ethnopharmacology,* 103, pp. 25-35.

Sharma, P.S. and Singh, V. (1989). Ethnobotanical Studies in Northwest & Trans-Himalayas V. Ethno-veterinary Medicinal Plants Used in Jammu & Kashmir State, India. *Journal of Ethnopharmacology,* 27 (1-2) : 63-70.

Singh, B. and Chaurasia, O.P. (2000). Medicinal Flora of Indian Cold Desert. *Acta Horticulture,* ISHS, Pub. Belgium. 523(13) : 65-72.

Stewart, R.R. (1972-1976). Flora of West Pakistan. Edited by E. Nasir & S. I. Ali. *An Annotated Catalogue of the Vescular Plants of West Pakistan and Kashmir*. Fakhri Printing Press, Karachi.

Stobdan, T., Chaurasia, O.P., Korekar, G., Mundra, S., Ali, Z., Yadav, A., and Singh, S.B. (2009). Attributes of Seabuckthorn (*Hippophae rhamnoides*) to meet nutritional requirements in high altitudes. *Defence Science Journal*, Vol 60, No. 2 : 226-230.

Parimelazhagan, T., Chaurasia, O.P. and Ahmed, Z. (2008). Bioactive substances and theraputic uses of seabuckthorn oil.. In *Book "Seabuckthorn : Hippophae spp.* : The Golden Bush. Eds. Sanjai K Dwivedi *et al.* Satish Serial Publication, New Delhi pp. 177-195.

Stobdan, T., Chaurasia, O.P., Yadav, A. and Singh, S.B. (2008). Seabuckthorn (*Hippophae rhamnoides*): A wild shrub of defence and societal importance in cold aird high altitudes. In *Book: In the service of Soldier - Life Sciences Perspective*. Eds. Manas K Mandal. Macmillan, New Delhi. pp. 402-411.

□□□

Chapter-9

Phytochemical and Pharmacological Analysis of Rhodiola sp. : A Review

Raj Kumar, Amol Tayade, Ashish Kumar, O.P. Chaurasia and R.B. Srivastava
Defence Institute of High Altitude Research (DIHAR), DRDO, Leh, Ladakh

1. Introduction

The genus Rhodiola (Crassulaceae) consists of more than 100 species growing mainly in Asia and Europe. *Rhodiola* Sp. (Golden root) has historically been used as an adaptogenic in Russia, Northern Europe and in China as a traditional herbal medicine and valued for its ability to enhance human resistance to stress or fatigue and promote longevity (Spasov *et al.*, 2000; Tolonen *et al.*, 2003). Rhodiola grows at high altitude mountain areas where the ecological environment is vulnerable to unfavorable factors. Rhodiola is a rare genetic resource due to limitation of its ecological adaptation. *Rhodiola imbricata* (Roseroot, Artic root, or Shrolo), is an important food crop and medicinal plant in Trans-Himalayan cold desert. It is mainly known for its adaptogenic functions including central nervous system stimulating activity, optimization of cardiovascular functions, antidepressent, performance enhancing abilities, elimination of fatigue, immunomodulatory properties and to prevent high-altitude sickness

(Gupta *et al.*, 2008). Golden root phytochemical extracts are the source of important commercial preparations widely used throughout Europe, Asia and more recently in the USA, with biological activities including antiallergenic and anti-inflammatory effects, enhanced mental alertness and a variety of therapeutic applications (Tolonen *et al.*, 2003). While phenyl ethanoid derivatives including salidroside (rhodioloside) were previously used exclusively to standardize medicinal preparations of Rhodiola extracts, it is now believed that a variety of co-occurring phytochemical constituents in the plant including phenylpropanoids (e.g. rosavins) and possibly terpenoids and flavonoids may be responsible for its unique pharmacological activity (Brown *et al.*, 2002).

The phytochemical constituents in Rhodiola are species-dependent, with *R. rosea* being the predominant species screened in efficacy studies although salidroside content in other species including *R. sachalinensis*, *R. kirilowii*, and *R. crenulata* have also been reported (Wu *et al.*, 2003). Two central Asian *Rhodiola* species (*R. heterodonta* and *R. semenovii*), of the Crassulaceae family, have similar rich ethnobotanical histories and are locally valued as adaptogens. *R. heterodonta,* indigenous to Uzbekistan, has been found to contain salidroside in common with Golden root and its extract is included, along with extracts of other *Rhodiola* species, in a medicinal preparation (carpediol) used to treat depression (Wikman and Panossian 2003). *R. semenovii,* indigenous to Kyrgyzstan and locally valued for endurance-enhancing and anti-hypoxic properties, contains dimeric and oligomeric proanthocyanidins which have been linked to the hypolipidemic, hypocholesteremic and anti-inflammatory properties of extracts from this plant. *R. semenovii* was reclassified by some Russian authors as a separate genus (*Clementsia semenovii*) because it did not contain detectable levels of tyrosol, salidroside and herbacetin characteristic of other members of Rhodiola (Kurkin *et al.*, 1986).

In previous studies, complete phytochemical characterization and cross-comparison between various species has been complicated by the intermittent use of dried, powdered or pulverized plant preparations for analysis. Although the natural isomeric forms, functional groups and bioactivity of many phytochemical components may be altered during extraction and processing in many cases the methods used to prepare herbal extracts/dried preparations are unknown or vary widely between processors. The objectives of such studies were to compare the phytochemical constituents of the *Rhodiola* species (*R. heterodonta* and *R. semenovii*) to the well-characterized Golden root (*R. rosea*), using freshly harvested live rhizomes and to standardize reproducible extraction and separation methods, in order to examine the inherent constituents potentially responsible for the above mentioned metabolic-enhancing properties. All three species were characterized in terms of phenyl ethanoid or cyanogenic glycosides and proanthocyanidin content.

2. Traditional and Current Medical Uses of *Rhodiola*

Rhodiola sp. (Crassulaceae, syn. *Sedum rhodiola* - DC. *Sedum rosea*- (L.) known by the common names Rhodiola, Golden Root, Arctic Root) has long history as a valuable medicinal plant having appeared in the Materia Medica of number of European countries (Linne, 1749). The plant grows in crevices of mountain rocks and on sea cliffs of Arctic regions of Europe, Asia (mainly in Siberia) and N. America. *R. rosea* was in use long ago by the Vikings as a medicine (Dragland and Galambosi, 1996), but this is somewhat speculative. In Linne´s Materia medica (Linne, 1749) the root of the Rhodiola is recommended in the treatment of headaches, "hysteria", hernias, discharges and as an astringent. *R. imbricata* is a known medicinal plant in the Amchi system of medicine (Tibetan system of medicine) being used in various ailments like medicine against lung problems, cold, cough, fever, loss of energy and pulmonary complaints. It is also used in preparation of health tonic and restores memory (Chaurasia *et al.*, 2007).

As a dietary supplement, numerous preparations of Rhodiola extracts are used world-wide (Khanum *et al.*, 2005). The functional claim of Rhodiola dietary supplements currently mentioned in the consolidated list of Article13 health claims of the European Food Safety Authority (EFSA) is formulated as following-"contributes to optimal mental and cognitive activity". In Sweden, Rhodiola tablets containing *R. rosea* SHR-5 extract have been on the market since 1985. They are currently registered as Traditional Herbal Medicinal Products (THMP) indicated as an adaptogen (Box 1) in situations of decreased performance such as fatigue and sensation of weakness.

3. Phytochemistry of *Rhodiola*

Rhodiola sp. has received great interest for phytochemical investigation for many years and many chemical constituents of

Rhodiola rhizomes have been isolated and identified from this species including phenylpropanoides, flavonoids, sterols, glycosides, organic acids (oxalic, citric, malic, gallic, succinic), phenolics including tannins and proteins (Ma *et al.*, 2006; Ali *et al.*, 2008). The dried rhizomes contain 0.05 percent essential oil. More than 50 polar compounds have been isolated from the water alcoholic extracts, they are monoterpene alcohols, their glycosides and cyanogenic glycosides (Fig. 1).

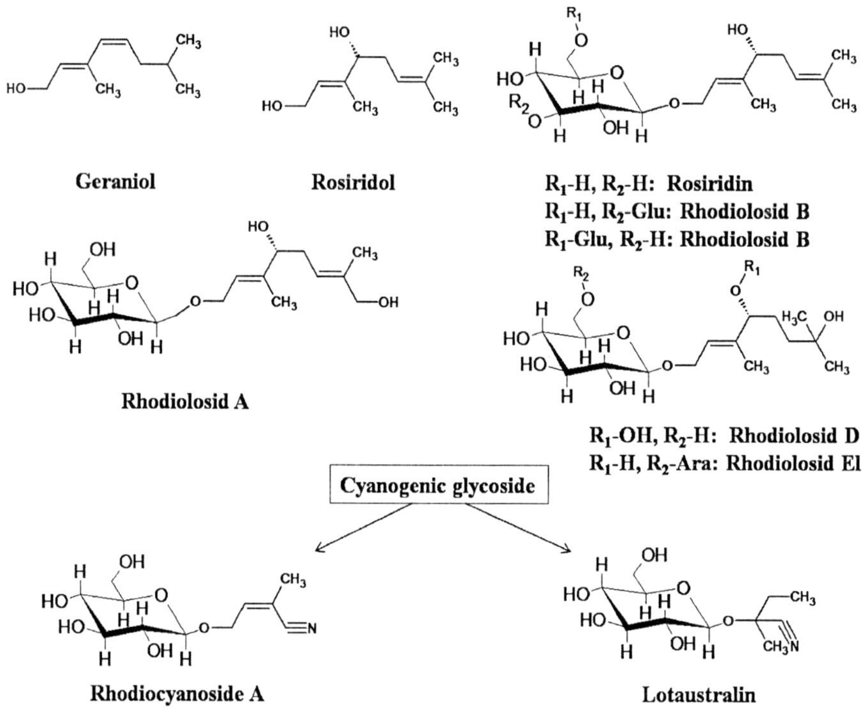

Fig. 1 : Glycosides and cyanogenic glycosides of *Rhodiola* sp.

The main chemical classes of volatile compounds were monoterpene hydrocarbons, monoterpene alcohols and straight chain aliphatic alcohols which accounted for more than 86 percent of the essential oil (Rohloff, 2002). The most abundant compound was decanol (30.38%). Other important constituents were geraniol (12.49%), 1,4-p-menthadien-7-ol (5.10%), limonene (4.90%), a-pinene (4.69%) and dodecanol (3.67%). Monoterpenes (geraniol, myrtenol, pinocarveol) as well as alipathic (heptanol) and aromatic alcohols (cinnamyl alcohol and cuminyl alcohol) were found to be the main compounds in Rhodiola essential oil. In contrast to results reported by a Russian research group the proposed main constituents such as octadecadienoic acid, heptanol derivates and hexadecanoic acid were only detected in insignificant amounts in the Norwegian samples (Belov *et al.,* 1994).

4. Total Phenol and Flavonoids in *Rhodiola*

Phenols are ubiquitous secondary metabolites in plants and possess a wide range of therapeutic uses such as antioxidant, antimutagenic, anticarcinogenic, free radical scavenging activities and also decrease cardiovascular complications. The scavenging ability of the phenols is mainly due to the presence of hydroxyl groups. Comparison of total phenol content (Table 1) between *Rhodiola* sp. shows significant variation in each extract. Total phenol content in *R. heterodanta* root extract is higher than shoot extract. *R. heterodanta* root extract with total phenol contents of 79.21 ± 0.26 mg GAE/g plant extract had the highest amount among the plant extracts analyzed in this study (Kumar *et al.,* 2010). *R. heterodanta* root and shoot extracts contained higher total flavonoid content in comparison to *R. imbricata.* In case of flavonoid concentration in *R. heterodanta,* it was found that root extract presented the highest values (269.3 ± .82 mg Qc/g plant extract).

Table 1 : Total phenol and flavonoids quantity of *Rhodiola* sp.

Sample	(Hydro-alcoholic extraction)	
	Total phenols (mg GAE/g plant extract)	**Total flavonoids (mg Qc/g plant extract)**
R. heterodanta		
Root	79.2 ± .0.24	269.3 ± 0..82
Shoot	26.16 ± .0.42	185.2 ± 1.13
R. imbricata		
Root	49.8 ± 1.37	167.4 ± 0.71
Shoot	21.24 ± 0.82	151.6 ± 1.31

Values are mean ± S.E.M (*n*=3)

5. Pharmacological Activities

Pharmacological effects of Rhodiola extracts are summarized below:

- Adaptogenic and stress-protective (neuro-cardio and hepato-protective)

- Cardioprotective
- Antioxidant
- Stimulating effect on the central nervous system including effects on cognitive functions such as attention, memory and learning
- Anti-fatigue
- Anti-depressive and anxiolytic
- Normalizing Endocrine Activity
- Increased Life Span

Rhodiola imbricata Edgew, growing on rocky slopes of western Himalayas (4000–5000 m), was found to possess radio-protective (Arora *et al.*, 2005), cytoprotective, antioxidant (Kanupriya *et al.*, 2005), wound healing (Gupta *et al.*, 2007) and immunomodulatory (Mishra *et al.*, 2006) activities. In a dose dependent study an aqueous extract of *R. imbricata* roots were found to possess potent adaptogenic activity in animals at a dose of 100 mg/kg body weight, during exposure to multiple stressful conditions of cold (5°C)-hypoxia (428 mm Hg)-restraint (C-H-R) and post stress recovery. The extract was also found to be safe in acute (LD_{50} > 10 g/kg) and sub-acute studies (Gupta *et al.* 2008). The studies on aqueous extract of *R. imbricata* root was also reported to aneliorate heavy metal toxicity (Saggu *et al.*, 2006). The effect of Rhodiola dry root vacuum dried extract treatment was studied on circulating energy fuels viz. serum glucose, serum lactate and serum NEFA, stored glycogen reserves and status of some of the key metabolic regulatory enzymes hexokinase (HK), phospho-fructokinase (PFK), citrate synthase (CS) and glucose-6-phosphate dehydrogenase (G-6-PD) in blood, liver and gastrocnemius muscle of rats on attaining the T_{rec}23°C during exposure to C-H-R and after recovery from C-H-R induced hypothermia (T_{rec}37°C) (Ramachandran *et al.*, 1990).

6. Antioxidant Activity

The DPPH scavenging activity of the hydro-alcoholic extracts from *R. imbricata* and *R. heterodanta* are given in (Table 2). The *R. imbricata* hydro-alcoholic root extract showed the highest scavenging activity. Free radical scavenging activity increased with an increasing concentration. The ability of root extracts to scavenge DPPH radicals suggests that they are electron donor and can react with free radicals to convert them to more stable products and terminate radical chain reactions.

Table 2 : DPPH radical scavenging activity of Hydro-alcoholic Extracts of *Rhodiola* sp.

Conc. (μg/ml)	Ascorbic Acid	*R.imbricata* Root	*R.heterodanta* Root	*R.imbricata* shoot	*R.heterodanta* shoot
100	36.55±0.61	32.06±0.68	32.34±1.53	11.20±0.65	10.21±0.85
200	39.11±0.80	36.78±1.76	35.46±0.65	12.05±0.88	12.05±0.65
400	41.26±0.83	40.11±1.10	39.43±0.88	13.75±0.88	13.04±0.65
600	43.14±2.82	42.98±0.86	41.84±0.65	14.04±1.12	13.90±0.65
800	46.90±1.01	45.51±0.68	45.39±0.65	16.02±0.88	14.89±0.42
1000	49.19±1.06	48.39±0.71	46.80±0.85	17.16±0.65	16.31±0.65

Values are mean ± S.E.M (n=3)

The reducing power of *Rhodiola* sp. is shown in (Table 3). *Rhodiola* sp. consists of hydrophilic polyphenolic compounds that cause greater reducing power. It is increased as the extract concentration increases. At different concentrations (100–1000 μg/ml) *R. imbricata* and *R. heterodanta* demonstrated powerful reducing capacity and these differences were statistically significant ($p<0.05$). *R. imbricata* hydro-alcoholic root extract showed the highest reducing power (Kumar *et al.* 2010).

Table 3 : Reducing power by Hydro-alcoholic Extracts of *Rhodiola* sp.

Conc. (μg/ml)	Ascorbic Acid	*R.imbricata* Root	*R.heterodanta* Root	*R.imbricata* shoot	*R.heterodanta* shoot
100	0.30±0.010	0.16±0.001	0.03±0.002	0.03±0.002	0.09±0.004
200	0.57±0.002	0.35±0.003	0.07±0.005	0.04±0.004	0.12±0.005
400	1.01±0.002	0.88±0.005	0.46±0.003	0.05±0.009	0.13±0.005
600	1.81±0.004	1.71±0.006	0.53±0.003	0.09±0.002	0.15±0.004
800	2.31±0.004	2.09±0.005	0.88±0.006	0.09±0.001	0.20±0.003
1000	2.67±0.004	2.41±0.004	1.04±0.004	0.13±0.004	0.22±0.005

Values are mean ± S.E.M (n=3)

7. Conclusion

Rhodiola sp. are popular plants in traditional medical systems in the Nordic countries, Eastern Europe and Asia, with a reputation for stimulating the nervous system, decreasing depressive traits, enhancing work performance, eliminating fatigue and preventing high altitude sickness. The traditional medicinal use of the plant, in addition to modern clinical use as referenced in scientific publications and official pharmacopoeias contribute to substantiate it's well-established medicinal use. These results can be useful for further application of Rhodiola sp. or its constituents in pharmaceutical preparations. With these kinds of investigation it would be easier to treat and prevent the cellular damage occurring due to free radicals.

References

Ali, Z., Fronczek, F.R. and Khan, I.A. (2008). Phenylalkanoids and monoterpene analogues from the roots of *Rhodiola rosea*. *Planta Medica*, 74: 178–181.

Arora, R., Chawla, R., Sagar, R., Prasad, J., Sing, S., Kumar, R., Sharma, A., Singh, S. and Sharma, R.K. (2005). Evaluation of radioprotective activities of *Rhodiola imbricata* Edgew - A high altitude plant. *Molecular Cellular Biochemistry*, 273: 209–223.

Belov, V.N., Lavrova, T.V., Vashkevich, N.G. and Mikhailov, A.Y. (1994). Extraction of essential oils from plant raw material by steam distillation. *Russian Journal Applied Chemistry*, 67: 154–156.

Brown, R.P., Gerbarg, P. and Ramazanov, Z. (2002). *Rhodiola rosea*: A phyto-medicinal overview. *Herbal Gram*, 56: 40-52.

Chaurasia, O.P., Ahmed Z. and Ballabh, B. (2007). Ethnobotany & Plants of Trans Himalayas, Satish Serial Publishing House, Delhi.

Dragland, S. and Galambosi, B. (1996). *Medisinplanter, Rosenrot, Forskningsparkeni°* As., pp. 143–145.

Gupta, A., Kumar, R., Upadhayay, N.K., Pal, K., Kumar, R. and Sawhney, R.C. (2007). Effects of *Rhodiola imbricata* on dermal wound healing. *Planta Medica*, 73: 774–777.

Gupta, V., Saggu, S., Tulsawani, R.K., Sawhney, R.C. and Kumar, R. (2008). A dose dependent adaptogenic and safety evaluation of *Rhodiola imbricata* Edgew, a high altitude rhizome. *Food Chemical Toxicology*, 46: 1645–1652.

Kanupriya, Prasad, D., Sairam, M., Kumar, R., Sawhney, R.C., Sharma, S.K., Illavazhagan, G., Kumar, D. and Banerjee, P. K. (2005). Cytoprotective and antioxidant activity of *Rhodiola imbricata* against tert-butyl hydroperoxide induced oxidative injury in U-937 human macrophage. *Molecular Cellular Biochemistry*, 275: 1–6.

Khanum, F., Bawa, A.S. and Singh, B. (2005). *Rhodiola rosea*: A versatile adaptogen, comprehensive Review. *Food Science Food Safety*, 4: 55–62.

Kumar, R., Tayade, A., Chaurasia, O.P., Hota, S. and Bala, S. S., (2010). Evaluation of anti-oxidant activities and total phenol and flavonoid content of the hydro- alcoholic extracts of *Rhodiola* sp. *Pharmacognosy Journal*, 2(11): 431-435.

Kurkin, V.A. and Zapesochnaya, G.G. (1986). Chemical composition and pharmacological properties of *Rhodiola rosea. Chemical and Pharmaceutical Journal*, 20: 1231-1244.

Linne , C. (1749). Materia Medica. LiberI.de Plantis, Stockholm, p.168.

Ma,G., Li,W., Dou, D., Chang, X., Bai, H., Satou, T., Li, J., Sun, D., Kang, T., Nikaido, T. and Koike, K. (2006). RhodiolosidesA-E, monoterpeneglycosides from *Rhodiola rosea. Chemical Pharmaceutical bulletin*, 54: 1229–1233.

Mishra, K.P., Padwad, Y.S., Jain, M., Karan, D., Ganju, L. and Sawhney, R.C. (2006). Aqueous extract of *Rhodiola imbricata* rhizome stimulates proinflammatory mediators via phosphorylated 1 kB and transcription factor nuclear factor-kB. *Immunopharmacology Immunotoxicology*, 28: 201–212.

Ramachandran, U., Divekar, H.M., Grover, S.K. and Srivastava, K.K. (1990). New experimental model for evaluation of adaptogenic products. *Journal Ethnopharmacol*ogy, 29: 275–281.

Rohloff, J. (2002).Volatiles from rhizomes of *Rhodiola rosea* L. *Phytochemistry*, 59: 655–661.

Saggu, S., Gupta, V., Sawhney, R.C., Rai, P.K. and Kumar, R. (2006). Analysis of heavy metal in herbal extracts of high altitude growing plants *Hippophae rhamnoides* and *Rhodiola imbricata. Toxicology International*, 13: 111–117.

Spasov, A.A., Mandrikov, V.B. and Mironova, I.A. (2000). The effect of the preparation rodiosin on the psychophysiological and physical adaptation of students to an academic load. *Eksp Klin Farmakol*, 63: 76-78.

Tolonen, A., Hohtola, A. and Jalonen, J. (2003). Comparison of electrospray ionization and atmospheric pressure chemical ionization techniques in the analysis of the main constituents from *Rhodiola rosea* extracts by liquid chromatography/mass spectroscopy. *Journal Mass Spectroscopy*, 38: 845-853.

Wu, S., Zu, Y. and Wu, M. (2003). High yield production of salidroside in the suspension culture of *Rhodiola sachalinensis. Journal of Biotechnology*, 106: 33–43.

Wikman, G. and Panossian, A. (2003). *Medicinal herbal extract carpediol for treating depression. U.S. Pat. Appl. Publ.*

□□□

Chapter-10

Soil Health Status of Trans-Himalayas

Somen Acharya, Narendra Singh, Anand K. Katiyar, S.B. Maurya and R.B. Srivastava
Defence Institute of High Altitude Research (DIHAR), DRDO, Leh, Ladakh

1. Introduction

Soil is a living system that represents a finite resource vital to life on the earth. It forms a thin layer of unconsolidated minerals and organic matter on the earth's surface that performs many processes essential to life. It develops slowly from parent materials and is modified by various soil forming factors like time, climate, macro and microorganisms, vegetation and topography. Soils are complex mixtures of minerals, organic compounds and living organisms that interact continuously in response to natural and imposed biological, chemical and physical activities. It serves as a substrate for plant growth, as a nutrient reservoir and as the site for many biological processes involved in decomposition and recycling of plant and animal residues (Wienhold *et al.,* 2004). People are dependent on soil and conversely, fertile soils are dependent on people and the use they make of the land. Fertile soils provide starting point for successful agriculture (Karlen *et al.,* 2001).

2. Soil Fertility

Soil fertility is a complex quality of soils that is closest to plant nutrient management. It is the component of overall soil productivity that deals with its available nutrient status and its ability to provide nutrients out of its own reserves and through external applications for crop production. It combines several soil properties (biological, chemical and physical), all of which affect directly or indirectly nutrient dynamics and availability. Soil fertility is a manageable soil property and its management is of utmost importance for optimizing crop nutrition on both short-term and long-term basis to achieve sustainable crop production. Soil productivity is the ability of a soil to support crop production determined by the entire spectrum of its physical, chemical and biological attributes. Soil fertility is only one aspect of soil productivity but it is a very important one. For example, a soil may be very fertile but produce only little vegetation because of lack of water or unfavourable temperature. Even under suitable climatic conditions, soils vary in their capacity to create a suitable environment for plant roots. For the farmer, the decisive property of soils is their chemical fertility and physical condition, which determines their potential to produce crops. Good natural or improved soil fertility is essential for successful cropping. It is the foundation on which all input-based high-production systems can be built.

3. General Fertility of Desert Soil

Fertility of desert soil is generally low with very less amount of organic carbon in it. Desert soils often have very low level of nitrate nitrogen and available phosphorus. The surface soils of the desert usually have the highest level of nutrients and make the soil very susceptible to degradation. Top soil removal by land use change, water and wind erosion can remove a large portion of soil nutrients. Most desert soil is basic in nature with pH often varying from 8 to 10. This high pH can further limit the availability of many nutrients. Organic carbon is very limited in desert soil rarely reaching 1 percent and often found below 0.3 percent. Adding organic matter often improves degraded desert soils.

4. Soil Health Status of Cold Desert Ladakh Region

Ladakh is the northern most part of India. It constitutes over 80 per cent of the Trans-Himalayan tract in India and is home to a unique assemblage of flora and fauna in its high altitude wetlands.

The topography is rugged with rocky terrains (Acharya *et al.*, 2010). Valleys are sandy, open, long broad with some water bodies. Indus, Shyok, Zanskar, Nubra, Suru and Dras are the main rivers of this region. The region, popularly known as a cold desert, is characterized by severe arid conditions. This Trans-Himalayan land is an icy barren landscape. Glistering white snow capped peaks and glaciers dominate in this most inhospitable terrain of its kind in the world. Temperatures may drop to -40°C in the long winter months between December to March and may rise to 35°C in the short summer season from May to August. Ladakh remains cutoff from rest of the world by surface transportation during winter months near about six months.

For determination of soil health status of this cold arid region, Defence Institute of High Altitude Research (DIHAR) had surveyed and collected soil samples from several villages of Leh and Kargil region. Depending upon the land holding, 3-6 samples were collected from each village. The samples were analyzed for various physico-chemical parameters which are indicator of soil health of a particular region. The soil analysis data (Table 1) revealed that the type of soil in the region depends upon weathered rocks in the particular location and found sandy to sandy loam in nature.

Table 1 : Range of chemical properties of soils of different villages of Leh and Kargil district

Parameters	Leh		Kargil	
	Minimum	Maximum	Minimum	Maximum
pH (ds/m)	5.65	10.12	6.57	9.47
Electrical Conductivity (EC) ds/cm	0.05	1.56	0.08	1.55
Organic carbon (%)	0.01	1.41	0.21	3.2
Phosphorus (Kg/ha)	4.0	25.66	2.80	116.1
Potassium (kg/ha)	12.32	496.15	115.0	861.3
Zinc (ppm)	0.06	6.0	0.14	5.17
Iron (ppm)	0.54	33.79	0.20	38.1
Copper (ppm)	0.43	3.52	0.48	3.41
Manganese (ppm)	0.2	32.17	0.40	4.74

Majority of the soils are sandy to sandy loam in texture and low in organic matter with poor water holding capacity. The pH ranged from 5.65 to 10.12 in Leh and pH of Kargil varied from 6.57-9.47. More than 90 percent samples from both the districts were having pH in the ranges of 7-9. Therefore, majority of the soils in high altitude cold desert regions are alkaline in nature. Electrical conductivity ranged from 0.05-1.56 dS/m in Leh and 0.08-1.55 dS/m in Kargil, respectively. More than 90 percent soils were low in available phosphorus but high in potassium with availability of the potassium depending upon the other parameters of the soil. Micro nutrients status like Zn, Mn Cu and Fe were also observed low in most of the collected samples.

5. Soil Health Status of Ladakh Region Before and After Cloudburst

Climate change is an age old phenomenon and has now emerged as one of the most burning environmental issues ever to confront the humanity. It has resulted in several kinds of anomaly in weather phenomenon all over the world. Recent unprecedented cloudburst in cold desert Ladakh leading to flash floods in several villages was one of such incident. On 5th August, 2010 night, suddenly cloudburst occurred in cold desert Trans-Himalayan region of Leh-Ladakh. Leh-Ladakh is an unusual place for a cloudburst because it is a cold desert region where average rain fall is very low. The maximum ever recorded rainfall in Leh was 96.2 mm in a 24 hour period measured in 1933. This cloudburst yielded 250 mm rainfall within half an hour resulted in a severe flash flood. Due to flash floods resulted cloudburst, 28 thousand quintals of food grains of different standing crops spread over 1400 ha of agriculture land have been damaged by severe soil erosion and deposition of boulders along with debris.

After flash flood devastated several villages of Ladakh region like Choglamsar, Saboo, Phyang, Shey, Nimoo etc.; soil samples from those villages were collected. Sampling was done for both affected and unaffected fields. During sampling some visible effect of flash flood were observed. They are (a) washing of top soil which resulted in undulated topography of cultivated fields, (b) accumulation of boulders and stones, (c) damage to irrigation channels and field boundaries, (d) damage to fruit and forest trees and in some cases lands become totally unsuitable for cultivation (Fig. 1).

Unusual changes in some important physico-chemical parameters of soils have been observed in Leh-Ladakh before and after flash flood in August, 2010. In most of the affected fields, top fertile soil has been completely removed and the fields were full of boulders and debris. One important reason for severity of soil loss is texture of the soil of this region. In Ladakh, soil is mainly sandy to sandy loam in nature which is very susceptible to erosion.

Table 2 : Comparison of physico-chemical properties of soils of unaffected and affected fields of Ladakh region before and after flash flood

Parameters	Before	After
pH	8.02	8.66
EC (ds/cm)	0.34	0.38
OC (%)	1.09	0.24
N (Kg/ha)	388.9	115.1
P (Kg/ha)	7.9	1.0
K (Kg/ha)	284.1	103.4
Zn (ppm)	0.17	0.09
Fe (ppm)	1.45	0.66
Cu (ppm)	0.67	0.21
Mn (ppm)	1.4	0.27
Sand (%)	57	73
Silt (%)	24	16
Clay (%)	19	11
*WHC (%)	47	19

* Water holding capacity

Soil pH was found increased from slightly alkaline to highly alkaline (8.66) in nature. However range of EC after flash flood was not significantly increased (0.34 to 0.38 mmhos/cm). As top soil was completely washed away, Organic Carbon (OC) as well as all macro (N, P and K) and micronutrients were severely reduced as compared to original soil. OC content is reduced to 0.24 percent from 1.09 percent. N, P and K content was recorded 115.4, 1.0 and 103.4 kg/ha respectively. Whereas micronutrient status (Zn, Fe, Cu and Mn) of flood affected soils were found to be 0.09, 0.66, 0.21 and 0.27 ppm only. Particle size distribution showed, nature of soil changed from sandy-loam to sandy soil. Sand content of affected fields increased from 57 percent to 73 percent whereas clay content decreased from 19 percent to only 11 percent. As a result Water Holding Capacity (WHC) of the soils was also reduced from 47 percent to merely 19 percent. Soil came from hills deposited and formed very hard pan over the surface soil. The deposition comprising mostly of sand and silt, has caused damage to the standing crops in some areas. Difference of soil fertility status of Ladakh region before and after cloudburst is shown in Table 2. From overall result it is observed that soil fertility status of flood affected fields was significantly reduced and macro and micronutrient deficiencies were also observed.

Fig. 1(a&b) **:** Pictures showing cloudburst at Leh-Ladakh in August, 2010 which destroyed agricultural fields at different places of Ladakh region.

5.1 Ameliorative Measures for Flood Affected Areas

After the calamity, it is very difficult to grow any crop in affected fields without soil amelioration. For reclamation of flood affected soils, following recommendations may be adopted.

- In extremely damaged fields where it became undulating topography and full of stones and boulders, it is very difficult to ameliorate, so if possible compensatory land can be allotted to the affected farmers.
- In moderately affected fields, removal of big stones, filling with soils, deep ploughing and leveling of the fields have to be carried out. The fields where hard pan has formed need removal of the deposited hard pan, deep ploughing to mix surface and sub-surface soil have been recommended.
- Construction of field boundaries and repairing of irrigation channels is necessary. Along the field boundaries planting of trees like willow and poplar which are dominant tree species of this cold desert region are also recommended.
- As soils became completely sandy in nature, heavy dose of organic matter in the form of FYM, vermicompost etc. have to be added

to enhance soil structure, aggregation, soil WHC, aeration as well as incorporation of major nutrients into the soil.

- Compansation to the affected farmers may be analysised on the basis of nature of damage.

6. Measures for Improving Soil Health

6.1 Soil Erosion and Degradation Control Measures

Since the topography of the area is rugged, undulated with sloppy terrain and being erosion prone, soil erosion is accelerated especially during summer months when snow melts, resulting in heavy discharge in drains, washing away the prime agriculture lands, plantations and pasture areas. Increased soil erosion results into loss of top fertile soil forcing reduction in the crop productivity. Over one-third of the area in Himalayan region suffers due to soil erosion and degradation (Table 3).

Table 3 : Soil degradation (ha) in Himalayan region

State	Water erosion	Wind erosion	Physical deterioration	Complex problem	Total degraded area	Total geographical area
Jammu & Kashmir	5460	1360	200	--	7020 (31)*	22224
Himachal Pradesh	2875	--	1303	--	4172 (75)	5567
Uttarakhand	1554	--	2280	--	3834 (72)	5348
Sikkim	235	--	--	--	235 (33)	710
Arunachal Pradesh	4327	--	176	--	4503 (54)	8374
Mizoram	1187	--	--	694	1881 (89)	2108
Manipur	133	--	111	708	952 (42)	2233
Nagaland	390	--	--	605	995 (60)	1658
Tripura	425	--	203	--	628 (60)	1049
Meghalaya	1168	--	146	34	1208 (53)	2243
Total	17754	1360	4279	2041	25434 (49)	51514

*Figures in perentheses are percent of total geographical area
Source: NBSS&LUP (2004)

The north-western hills of Jammu and Kashmir, Himachal Pradesh Uttarakhand and hills of north-eastern states suffer due to severe soil erosion of more than 20 t/ha/year (Singh *et al.*, 1997). The adoption of appropriate soil and water conservation measures is essential for protecting the lands from soil erosion and deterioration in soil health. A number of mechanical and cultural management practices like land leveling, contour bunding, contour trenching, bench terracing, contour farming, intercropping, strip cropping, mixed cropping, mulching, crop geometry and vegetative barriers etc. are recommended for checking soil and water loss from the sloping lands. The runoff water could be harvested and stored in suitable storage structures for supplemental irrigations during moisture stress periods for the crops. The yield increases are over 100-200 percent even with one supplemental irrigation of 5 cm (Samra and Narain, 1998). In order to combat with this problem, in recent years series of gabion structures and check dams are laid to stop erosion in terms of bank cutting, scouring etc. at different vulnerable sites of this cold arid region (Fig. 2).

(a)

(b)

Fig. 2 (a&b) **:** Soil erosion control measures in different places of Ladakh

6.2 Adopting Integrated Nutrient Supply System

The soils of cold desert Ladakh region are generally deficient in nutrients like nitrogen, phosphorus, calcium, magnesium, sulphur, zinc, boron, molybdenum etc and are low to medium in organic carbon and K content. These need to be supplied by addition of inorganic and organic fertilizers to maintain steady supply of essential nutrients to crops. But, the supplies are seldom met by the farmers by applying very little of fertilizers. The beneficial effects of balanced and integrated nutrient management on soil health in terms of physical, chemical and biological health and overall crop productivity have been very well demonstrated in various parts of India. The conjunctive use of chemical fertilizers and organic manure (NPK+FYM) enhanced organic carbon, available nutrients, soil aggregation, water infiltrability, microbial biomass carbon and microbial population as compared to use of chemical fertilizers alone. The usefulness of integrated nutrient management has also been demonstrated in improving soil health and productivity of important vegetables in cold desert/dry temperate areas (Parmar, 2005). The combined use of major and micronutrient fertilizers, bio-fertilizers and FYM will help in obtaining quite higher yield of different vegetables like peas, potato, tomato, cauliflower, cabbage tomato, etc. in this cold desert region. The integrated nutrient management also improves quality of produce in terms of protein, starch, total soluble sugars and vitamins.

The integrated nutrient supply system which envisage conjunctive use of chemical and organic fertilizers therefore, is the most ideal system of nutrient management. The system enhances nutrient-use efficiency, maintains soil health, enhances yield and reduces cost of cultivation. There is need to augment the supplies of organic manures like FYM, green manure, compost/vermicompost etc. The use of biofertilizers is still minimal in this hilly and mountainous region and requires to be promoted by producing effective strains with enhanced shelf life. A variety of biofertilizers that could be popularized are nitrogen fixers (*Rhizobium, Azotobacter, Azospirillum*), Phosphate Solubilizing Bacteria (PSB), blue-green algae, mycorrhizae and plant growth promoting rhizobacteria (PGPR).

6.3 Soil Testing Facilities

Non availability of proper soil testing facility remains major constraint in Ladakh region. To overcome this problem DIHAR has established soil testing facility for testing of pH, EC, OC, N, P and K. The State Agricultural Departments should also take some intervention in this regard. Preparation of georeferenced soil fertility maps at district and block levels are necessary for precise fertilizer recommendations. There is an urgent need to open up some soil testing laboratories, at least one each in different blocks of this cold desert region for more coverage and enhanced capacity.

7. Conclusion

In near future, proper soil and water conservation measures should be adopted for this cold desert region for improving soil health and sustainable crop production for a long period of time. The soils of the entire region are almost sandy in nature and the irrigation net work is very rare. The vegetation is entirely dependent on irrigation. More irrigation canals need to be constructed to conserve water for better utilization as per site requirement. The canals should be treated through different soil conservation techniques to ensure better utility for crop production, afforestation and pasture development. For natural resource conservation, detailed village wise farm level database should be generated on priority basis. Availability of such a database will have an immense utility in undertaking timely restoration and conservation measures in these areas.

References

Acharya, S., Stobdan, T. and Singh, S.B. (2010). Seabuckthorn (*Hippophae* sp. L.): New crop opportunity for biodiversity conservation in cold arid Trans Himalayas. *Journal of Soil and Water Conservation,* 9(3): 201-204.

Karlen, D.L., Andrews, S.S. and Doran, J.W. (2001). Soil quality, current concepts and applications. *Advance Agronomy,* 74: 1-40.

NBSS&LUP (2004). Soil Resource Management Reports. National Bureau of Soil Survey and Land Use Planning. Nagpur, India.

Parmar, D.K. (2005). Increasing Vegetable Production in a Sustainable Manner through Integrated Use Of Major Nutrients, Micronutrients, Bio- Fertilizers and Organic Manure in Cold Desert And Dry Temperate Areas of Himachal Pradesh. *NATP Sub-Project Report*, CSKHPKV, Hill Agriculture Research & Extension Centre Bajaura, District, Kullu. 38p.

Samra, J.S. and Narain, P. (1998). Soil and Water Conservation. In Singh and Sharam (Ed) 50 Years of Natural Resource Management Research. NRM Division, ICAR, New Delhi.

Singh, G., Babu, R., Narain, P., Bhushan, L.S. and Abrol, I.P. (1997). Soil erosion rates in India. *Journal of Soil and Water Conservation,* 47(1): 97-99.

Wienhold, B.J., Andrews, S.S. and Karlen, D.L. (2004). Soil quality: a review of the science and experiences in the USA. *Environmental Geochemistry and Health,* 26: 89-95.

□□□

Chapter-11

Agricultural Extension and Technology Dissemination in Ladakh : An Overview

Dorjey Angchok, Pal Murugan M, A.K. Katiyar, Narender Singh, Shashi Prakash, Hitesh Kumar, Konchok Tamchos and R.B. Srivastava
Defence Institute of High Altitude Research (DIHAR), DRDO, Leh, Ladakh

1. Ladakh An Overview

The cold deserts of the Himalayas are a unique ecosystem lying in the Greater Himalayan ranges in the rain shadow of the greening monsoons further south. Altitudes of inhabited areas range from 8,500 to 16,500 ft. above MSL. Major precipitation is only in the form of snow. Temperatures (in inhabited areas) plummet to -40°C in winters and scale up to 35°C in summers. The Himalayan cold deserts include large pockets in the Western Indian Himalayas - the regions of Ladakh and northern Himachal Pradesh (Fig. 1).

In India, risk prone high altitude area constitutes the easternmost Trans-Himalayan part of Ladakh and Zanskar range (Jammu and Kashmir) and some parts of Lahul Spiti (Himachal Pradesh). Truly described as cold desert with low population density, Ladakh constitutes the easternmost Trans-Himalayan part of Jammu and Kashmir State of India comprising of two districts viz, Leh and Kargil, bordering Pakistan and China (Fig. 1). Geographical area of Leh covers an area of 82665 Sq.km and 51358 hectare of reporting area, situated along the valleys of the Indus river, with an estimated population of 117232 (Table 1).

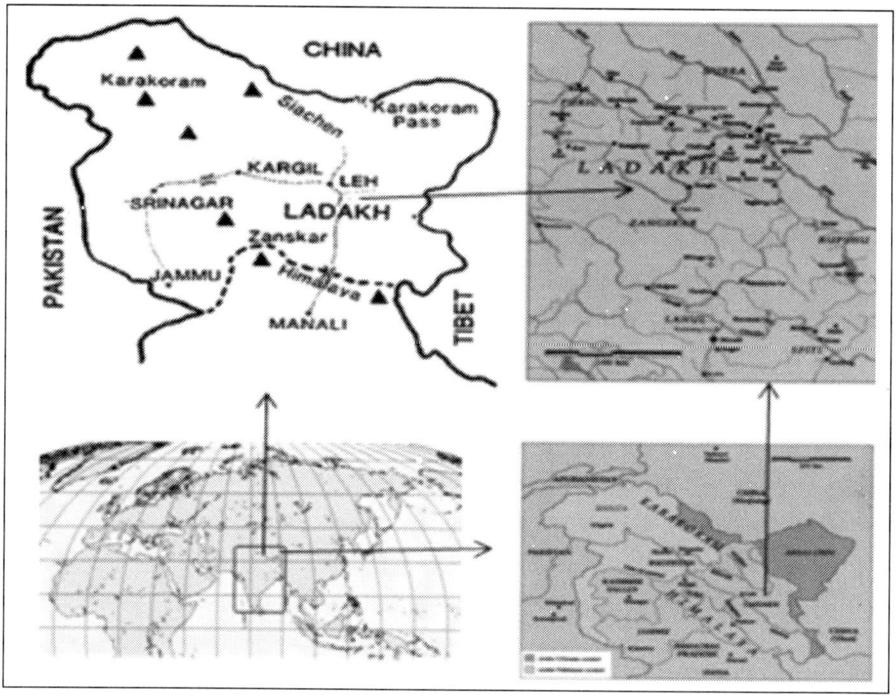

Fig. 1 : Geographic location of Ladakh

Description	Ref. Year	Unit	Magnitude
Geographical Area		Sq km.	82665
Reporting Area	2007-08	Hect.	51358
Number of Tehsils	2007-08	No.	03
Number of Blocks	2007-08	No.	09
Number of Villages	1981 Census	No.	114
Population	2001 Census	No.	117232
Households (Hh)	2001 Census	No.	24147
Average Hh. Size	2001 Census	No/Hh	05
Cultivators	2001	No.	22149
Net Area Sown	2007-08	Hect.	10193
Net Area Irrigated	2007-08	Hect.	10193
Gross Area Irrigated	2007-08	Hect.	10599

Table 1 : Leh at a glance

Source: Statistical Hand Book, Ladakh Autonomous Hill Development Council, Leh, 2007-08.

Leh is India's highest and one of its most arid, coldest and sparsely populated district. Geographically Leh district is sandwiched between two Himalayan ranges - Zanskar on its west and Ladakh on its east. Across the eastern range are the great highlands of Leh district - Chang Thang - which extend along the south of Ladakh and west, over the Indian-Chinese border, into Tibet. To the north of Leh district is the district of Kargil and beyond it is the Karakorum Range. The Indus River runs through Leh district from its south to its north, collecting the water of melting glaciers (Fig. 2) and straddling, several high-lying villages along its banks. Geo-ecologically Leh district is divided into almost halves - a cold, arid desert with minimal precipitation in the northern half, and high altitude grassland (Chang Thang) with perennial grasses in the southern half.

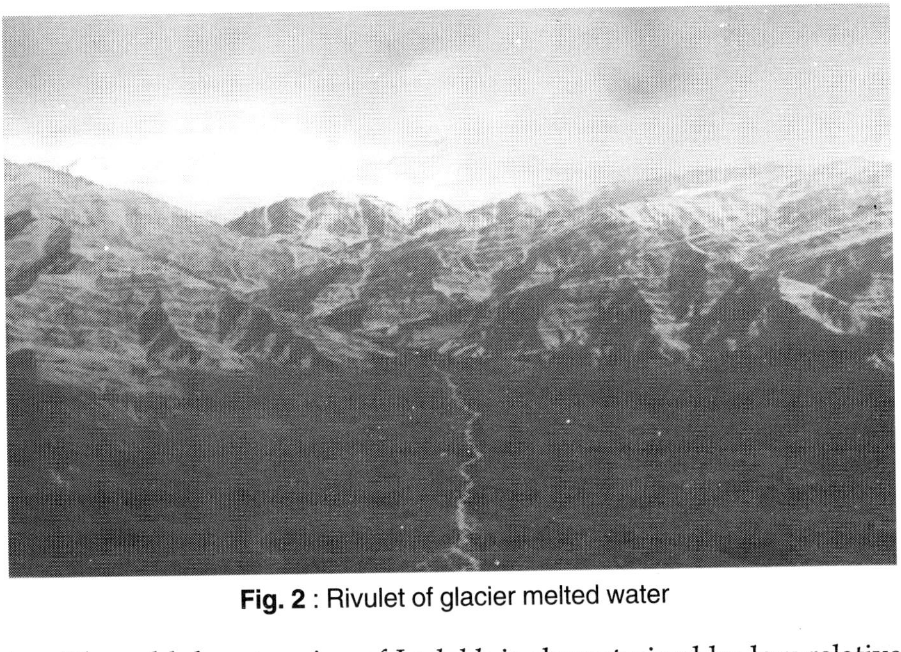

Fig. 2 : Rivulet of glacier melted water

The cold desert region of Ladakh is characterized by low relative humidity (20-40%), low atmospheric pressure (493 mm Hg), low partial pressure of oxygen, high wind velocity (8-10 km/hr), very low annual precipitation (80-300 mm), and sub-zero temperature (up to -40°C) during winter months. Intensive sunlight, high evaporation rate, strong winds and fluctuating temperature (35°C to -40°C) characterize the general climate. It is generally said that a man sitting in the sun with his feet in the shade can have sunstroke and frostbite at same time. With sparse vegetation there is little moisture in the atmosphere. Rains are very rare, though it may even snow during July-August, the hottest months. Because of high mountains all round and snowfall during winter, the area remains landlocked to the outside world for nearly six months in a year.

2. Agricultural Extension of DIHAR

From defence point of view, Ladakh is a strategically important region of India as it is having border areas with China and Pakistan; therefore there is a large chunk of defence forces deployed in this region. In this context to meet out the fresh food requirement of these deployed defence forces, Defence Institute of High Altitude Research (DIHAR) of Defence Research and Development Organisation (DRDO) was established in 1962 with a mandate to fulfill the fresh food requirement of defence personnel deployed in Ladakh sector of J&K State. DIHAR has core competence in cold-arid agro-animal technologies which it intends to transfer among the local farmers of this region through its extension activities.

DIHAR develops various agro-animal based technologies to be adopted by local farmers of Ladakh and finally the growers supply their produce to army through Cooperative Marketing Society. The agro-climatic condition of Ladakh is unique in a sense that here prevails high variation in temperature, high UV radiation, arid and sandy soil, low humidity and all these accompanied by high altitude makes it more difficult for production of fresh vegetables in this area.

In this context the various extension approach models adopted by DIHAR over the years since its establishment are given below

2.1 Early Phase

During the early period during 1960's and 70's there only few vegetables (like turnip, radish, carrot, potato etc.) being grown by local farmers for their own subsistence need. According to Reifenberg G, 1998, David Fraser, 1986 in his amusing account of travels around the turn of nineteenth century, found eggs, fowl and vegetables as part of the diet in *Lamayuru* (West of Ladakh). Then came the Moravian missionaries to Ladakh in the last quarter of the nineteenth century and they brought with them vegetables such as potatoes, spinach, cauliflowers, radishes, green beans, and tomatoes (Fraser D, 1986). These were probably the first significant additions to the Ladakhi diet for many years-turnips, swedes and peas having predominated until then. It was probably also the Moravians who demonstrated that root vegetables could be stored in pits in the ground (Rizvi J, 1983), thus improving the variety of foods available in the winter.

Regarding the concept of extension during those early periods, there were lack of formal extension strategies to spread the know-how and technologies among the masses. During those times any technology perceived as new (which may be available elsewhere but not available in Ladakh), if introduced in Ladakh then it was an innovation for Ladakh. Therefore, those introduced vegetables and their cultivation practices were an innovation for Ladakh. Due to lack of formal strategies these know-how remain limited to some of the aristocratic families only, who had good relations with the missionaries.

With the establishment of DIHAR, efforts were made to regularize and standardize the cultivation practices for vegetables. This was necessary because the agricultural practices followed elsewhere cannot be practised as such in Ladakh condition due to its different geographic and environmental condition. In this connection various trials were conducted at DIHAR campus Leh and it came out with certain recommended package of practices. But the extensions of these packages were a major hurdle because of its newness and lack of inputs availability like seeds to the farming community. Therefore during those times, major emphasis were given to production of quality seedlings at DIHAR campus Leh and then was distributed to the farmers. This effort was accompanied by training farmers regarding their cultivation practices.

Though this effort had a good impact but it was limited to only few nearby villages' only due to lack of manpower and logistic support.

2.2 Mid Phase

To reach out among the farming community at large it was necessary that DIHAR extends its activities in other parts of Ladakh also. Therefore detachments were established at Partapur in Nubra valley (145 kms approximately from Leh) in 1982 and Kargil in Suru valley (250 kms approximately from Leh) in1987.

During 1980 and 90's these new establishments also functioned well in their respective valleys with the back up from DIHAR, HQ Leh and mostly their work was to replicate the HQ model in their respective valleys. This way activity of DIHAR was extended into other areas also. For mass dissemination of information Kisan Mela was for the first time organized in 1984, which later on became a continued annual programme of DIHAR.

One major contribution during this period was the establishment of Farmer's Cooperative Marketing Society (CMS). It was necessary because the linkage between producer/growers (farmers) and consumers (defence forces) was weak and haphazard. The coming up of CMS helped defence forces to regularize the channel whereby they were able to negotiate the type and quantity of fresh vegetables CMS can supply to them.

Whatever limited extension activities have been carried out by DIHAR during those times were based on the classical diffusion model, where it is assumed that once a technology is diffused and adopted by a farmer (mostly progressive resource rich), the technology in a social system will diffuse of its own like a contagious disease. The major assumption is that it is a matter of time that all farmers will adopt the technology.

But in reality it is not the case and consequently we call those farmers (who have not adopted the technology) as laggards/ traditional/non-progressive/non-innovative/disadvantaged..., without assessing, whether the technology is compatible with their (farmers') needs, interest, culture, skill, and the resources available with them. In other words we find fault in the farmer but not in the technology.

2.3 Recent Phase

During the early 2000, DIHAR started to look into formal extension strategies. DIHAR has established extension Division in 2005. From then onwards extension as a separate wing started to operate. In the initial years of its establishment it was necessary to regularize the existing extension services of DIHAR and to bring out a model which is suited to the socio-economic, geographic and climatic condition of Ladakh region. In this context following initiatives were taken up:

2.3.1 Adoption process

a) *Awareness stage*: To make aware people about DIHAR technologies and reach out at the grassroots level, radio as mass media channel was/is being used. The reason being, radio as a mass media channel was found in a study to be the most preferred and credible source of information. As Ladakh being a large geographic area and villages settled in dispersed areas in various far flung areas where accessibility is a major problem, therefore, radio as a mass media channel was used effectively. In this regard experts of DIHAR regularly broadcast programme on various agro-animal technologies.

Apart from the above mentioned awareness program DIHAR has been organizing a two day "*Ladakhi Kisan Jawan Vigyan Mela*" regularly on the second Saturday and Sunday of August in Leh (since 1992), third week of August at the Det, Partapur (since 2002) and 1st week of September in Kargil to disseminate the latest agro-animal technologies. Various competitions are also organized for the army units and local farmers during the mela which is a unique collective venture of scientists, farmers and armed forces personnel to develop the difficult area of Ladakh. The Mela symbolizes the popular slogan 'Jai Jawan, Jai Kisan Jai Vigyan'.

b) *Interest stage*: After the awareness initiatives, farmers from various villages are provided training in which they are taught in detail the various aspects of technologies which they can adopt in their field condition. In this stage, they are provided with the basket of technologies among which they choose out of their interest and capabilities.

c) *Evaluation and trial stage*: After the farmer is interested in a particular technology, in this stage, farmer evaluates the respective technology at his own field condition. Various on-farm demonstrations are conducted at farmer's field under the supervision of DIHAR, where the farmers evaluate the technologies by themselves in their own field conditions.

d) *Assessment and refinement stage*: After the evaluation and trial stage, the farmers now assess the technology in terms of various expected inputs (both tangible and intangible) required, compatibility with his available resources, his needs, market opportunities etc. There also may be some cases where a farmer wants certain modification/changes in the technologies and these feedbacks are reported back to DIHAR for refinement of technologies.

e) *Adoption stage*: The famer after evaluation (mental and practical) decides to adopt the technologies. At this stage DIHAR provides the necessary inputs required (like seeds/seedlings) or make him aware of the various sources from where inputs can be availed. After adoption of the technology farmers remain in touch of DIHAR and give his feedback regarding the performance of the technologies.

In addition to the above mentioned stages, we need to look into the nature of the technology i.e. the appropriateness of the technology, whether it is in line with the needs (both felt and unfelt) of the farmer (both small and big). In this connection following measures can be taken up to improve upon such a situation:

- *Targeting approach*: Similar to the strategy adopted by the product marketing ventures, we have to approach and reach out the selected segment of the society for particular technology.
- Interest developed through tour exposure, field visits etc
- To increase the countervailing power of the farmers through capacity building.
- To involve local farmers at every stage of technology generation process.

2.3.2. Entrepreneurship development

In the last decade, DIHAR has actively worked into bringing about various agro-based products and has achieved to bring out products like herbal tea, seabuckthorn beverages, anti-oxidant supplement etc. All these products are based on plant resources available in Ladakh condition. It has been the endeavour of DIHAR to support local Self Help Groups (SHG's), local NGO's and local un-employed educated youth, to start their own business ventures based on these products and become successful entrepreneurs. In this context, following stages are given due consideration

a) *Stimulus*: To make youth aware of the basket of opportunities and help them select as per their capacity and needs.

b) *Support*: Capacity building and orientation about the potential enterprise.

c) *Sustenance*: To help them sustain in the already existing market. This is a very important phase because in most of the entrepreneurship programme the newly introduced entrepreneur is unable to handle and compete with the existing players in the market and finally he/she drops out.

2.3.3. Contract farming

In the recent past it has been felt that for a farmer to continue with agriculture has become a costly business and especially in Ladakh like condition it is more evident because discount rates/opportunity cost is quite high in other sectors like tourism. People (especially youths) do not prefer to work in their field because just working as a tourist potter or a guide or running a taxi fetch them more money relatively. Secondly, the readily available food material at subsidised rates through public distribution system (PDS) in almost every village adds on to the decreasing trend of people engaging into agriculture. Therefore to solve this perceived problem DIHAR has taken certain steps in which contract farming is one of them.

The concept of contract farming was introduced in Ladakh by DIHAR in the late 2000's in collaboration with PepsiCo private limited for the production of quality potato tubers and AVT Natural Product Limited for the production of marigold seeds. In the initial year of 2009 local farmers of Ladakh earned more than Rupees 75 lakh through contract farming. In the near future it is expected to diversify the crops for contract farming as per demand and resource available in Ladakh.

2.3.4. Capacity building

Since its establishment DIHAR has been regularly providing trainings to the local farmers of Ladakh. In the initial and mid phase it was necessary for DIHAR to take up an active role in providing training to the people because they were not aware of the available technologies.

But now the phase has arrived when everybody in Ladakh is aware of DIHAR and its activities. These days people (individual and groups) request for their training needs by themselves and DIHAR fulfils their needs. For the past few years most of the trainings were sponsored (by the trainee) programmes except for the people from remote far flung areas and defence personnel.

Following are some of the areas in which trainings are provided:

a) Vegetable nursery raising under protected and open condition
b) Protected vegetable cultivation techniques

c) Seed production of certain vegetables
d) Techniques of tree plantation and their management.
e) Cultivation and management of fruit orchard
f) Vermi-composting methods
g) Poultry rearing and management
h) Dairy farming management
i) Post-harvest management of vegetables and fruits
j) Cultivation of medicinal and aromatic plants

All these trainings are provided for which the duration, time and venue is decided as per convenience of the trainees. Apart from these, specific trainings (e.g. propagation techniques in apricot, apple, Seabuckthorn etc) are also provided as per the needs of the trainees and the programme is chalked out after having a pre informal discussion with the trainees.

2.3.5. Local knowledge system (LKS)

The term "Local Knowledge "refers to the knowledge of indigenous people as well as any other defined community. Local or traditional knowledge are the products of life long experience of the farmers, transferred from one generation to another. It is used synonymously with "Traditional" and Indigenous Knowledge ", to differentiate the knowledge developed by a given community from the International Knowledge System, sometimes called as "Western System", generated through universities, government research centers and private industries.

Only in the recent past, a growing number of scientists and academicians have started to analyze the consequences of their development efforts and have started giving due importance to the traditional wisdom which were/are prevalent on the earth past immemorial. In the words of Warren (1993), "developing countries have a valuable, but largely untapped reservoir of indigenous agricultural knowledge."

Taking into consideration these facts, in the recent past DIHAR has taken initiative to document Local Knowledge (LK) available in Ladakh region. Initially it was started in 2009 in collaboration with National Innovation Foundation (NIF) Ahmedabad. As NIF were the pioneers in this field initially their expertise were utilized after which

DIHAR initiated its own programme to tap this valuable resource. The basic objective to this programme is to conserve the local pool of knowledge into a kind of knowledge bank and then utilize it for generation of appropriate technologies and products after required value addition and scientific validation. With those technologies local farmers can identify themselves because it comes out of their own knowledge system.

2.3.6. Role of allied agencies

DIHAR being a R&D institution, always had limited manpower to carry out its extension activities because to reach out at the grassroots level and to achieve their confidence and establishing credibility among the farming community involves lots of frequent interactions (personal and impersonal both). After the establishment of more detachments in other valleys its need was felt more. Therefore DIHAR involved other allied agencies who are working in the similar field e.g. the local NGO's, women SHS's and state agriculture department. Now, except for whenever felt necessary in some rare cases DIHAR utilizes the organizational setup of these allied agencies to reach out at the masses. This helps in shortening the line of command and further adds to fast dissemination of information and feedback. In addition to this, it also helps in avoiding work duplication and confusion among the farmers.

2.3.7. Participation in exhibitions / Congress

DIHAR is one of the only institute working in the field of cold arid high altitude agro-animal sciences located at Ladakh region. Therefore to reach out to other outside agencies and partners it represent itself at various national and international forums like the DEFEXPO, India International Trade Fair (IITF), Science Congress and various other forums. Apart from spreading DIHAR information such activities help in two way flow of information and lot of learning experience is gained.

3. Learnings from DIHAR

3.1 High Altitude Community

Ladakh is characterized by it's unique socio-cultural and environment accompanied by harsh and inhospitable climatic conditions. In such a condition we cannot 'copy and paste' the various extension models (most of them developed for conditions prevailing in western developed countries) "as it is".

Similarly, as we have adopted in other disciplines, here in extension also we need to evolve our own strategies and action plan, appropriately suited to the conditions prevailing here in Ladakh.

A chronologically arranged activity is necessary for efficient execution of 'plan of work' and for success of any extension work being carried out by an organization. In this regard a conceptual framework is given below to help us acquaint with the basic outline in linking together the theoretical orientation and practical aspects of extension activities in Ladakh like condition.

3.2 Local Knowledge System (LKS)

The direction in which the western world is developing is not one which all must inevitably emulate. Western science generally do not find an explanation for many of the traditional and/or ancient practices our people, more specifically our farmers use, it does not on any way imply that the practices or technologies are not "scientific" (Angchok, 2000).

There is now concurrence developed among the so called scientific community also that LKS has great potential to learn from and integrate into the mainstream scientific domain. Therefore it is necessary that we incorporate LKS in our formal R&D network.

3.2.1 Incorporation of LKS in Research and extension network

There is the need for developing a framework for incorporating LKS into the Research and Extension System because:

(i) The transfer of technology (TOT) model fits badly with the needs and priorities of resource-poor farmers;

(ii) Agricultural extension programmes are still biased towards techniques and strategies, which are capital-intensive;

(iii) Resource-poor farmers (RPF) are scattered and are not able to make their needs and priorities readily known and felt; and

(iv) The TOT model cannot easily handle the complex interactions of RPF farming; links between crops, especially with intercropping and multiple tiers; agro-forestry and livestock-crop-tree complementaries; and the progressive adjustments required in the field in the face of seasonal and inter-annual fluctuations.

A framework for incorporating LKS system into agricultural research and extension should have following features:

(i) Strengthening the capacities of regional research and extension organizations.

(ii) Identifying the need for extension scientists/social scientists in an interdisciplinary regional research team.

(iii) Formation of a sustainable technology development consortium to bring farmers, researchers, NGOs and extension workers together well ahead of the process of technology development; and

(iv) Understanding that it is impractical to depend entirely on research stations for innovations, considering the inadequacy of human resource capacity of the regional research system.

In this connection Parasar (1994) developed a comprehensive model for incorporating local knowledge in Research and Extension Network (REN). The model can fit well for Ladakh like conditions also, with little or no modification. Major features of the suggested model are:

(i) To set up a Center for Indigenous Knowledge (CIK), assigning broadly at least three-fold work, viz., (a) collection, (b) documentation and (c) preservation of LK in a kind of Knowledge Bank;

(ii) Other activities of CIK should include coordination and monitoring, management of research outputs and related publications for developing literature on LKS and dissemination of center's activities;

(iii) Curricular sanction for local knowledge in the school and undergraduate courses and integration in the pre-service training courses of village level agricultural extension workers to remove attitudinal barrier against LKS;

(iv) Support of general academia in addition to agricultural disciplines;

(v) Involvement of NGOs in the process;

(vi) In order to popularize LKS among the scientists, extension functionaries as well as the farming communities, adequate scope for publicity through electronic and print media should be provided; and

(vii) The collected local agricultural knowledge should be subjected to preliminary screening and ready assessment in different professional forums before final recommendation.

3.3 Local Institution

Institution refers to the organised system of practices and social roles developed about a value or series of values, and the machinery evolved to regulate the practices and administer the rules. Studies have shown that more diverse the environment and lower the population density (as prevailing in Ladakh) greater is the need for social networking. Local institutions as a network platform play a major role in sustaining the life of people in such conditions by designing strategies to deal with the inhospitable and tough environmental conditions. Diversified uses of a natural resource are essential for household survival in high-risk areas and institutions for resource management are a part of evolutionary cultural, religious, and social experience of any community (Gupta 1991).

Keeping the above arguments in consideration, attempt has to be made to generate a framework to study local institutions (Fig. 3) in context of risk prone high altitude areas like Ladakh (for detail see Angchok, 2003).

Till date whatever limited initiatives have been taken up, it has been limited to tangible/material aspects of LKS e.g. use of medicinal plants. In this process we ignored the intangible/non-material part of LKS. The intangible part plays a major role in sustaining the life of local people in Ladakh e.g. the strong social-networking system and the presence of local institutions.

It is argued by different scholars that "institutions are like grammar, and technology the word" i.e. the technology has to fit within the institutional grammar, otherwise it has no meaning or may produce distorted meaning. In other words, institution to a great extent influences the nature of technology being generated and adopted.

There is need to identify the institutional diversity and infrastructure already existing/existed in Ladakh, and see how they can be effectively used for better extension of services from DIHAR.

Ladakh has a valuable, but largely untapped reservoir of Local Knowledge System (LKS), particularly in context of agriculture. These rich pool of LKS accompanied by strong social networking (in a high diversified and low population density) through local institutions help farmers in efficient management of their scarce resources in a sustainable manner. Therefore efforts should be made to use the prevailing local institutions for carrying out various extension activities.

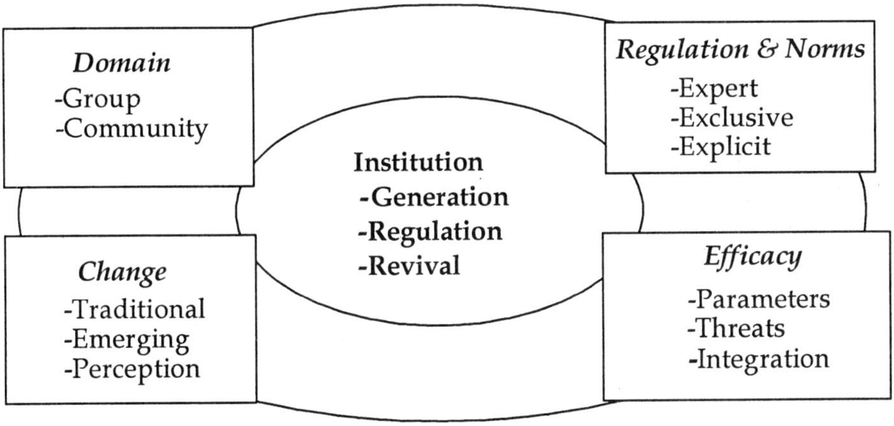

Fig. 3 : Framework to study local institution

4. Recommendations

4.1 Basic Data

Due to the mountainous physical environment in Ladakh access to villages is not easy. Therefore the importance of secondary data is immense to carry out an effective extension activities, for which we need to have all the necessary data available with us.

- Districts-blocks-villages (population, area, distance, socio-economic status, crops grown, and other specific features)
- Households (acreage, family size, major source of income, social status *etc.*).

Most of these data can be made available from various state departments and the Hill Development Council of Ladakh through formal request.

4.2 Identify Local Innovators

Local farmers are very ingenious and innovative in designing new technologies or practices to cope up with the environment. Though their innovations may be at initial (rudimentary) stage, but with the integration of proper modern scientific inputs into these innovations, the technology so developed can be made effective and viable (appropriate). e.g. A farmer from Gujarat got patent for his designing of a cotton stripper machine, with the help (modern scientific inputs) from CSIR, NID and NIF.

Similar programme can also be initiated in Ladakh region, through collaboration with various Defence Research & Development Organisation (DRDO) laboratories. It will not only help the organization but also the local farmers.

4.2.1. Viable network

According to Gupta (1997), a network should be established to link together the local innovator, institution, and enterprise. It is an instrument that will help those innovators whose knowledge, practices and innovation can be commercialized or disseminated by the state or NGOs among potential users. It tries to achieve market & technological linkage.

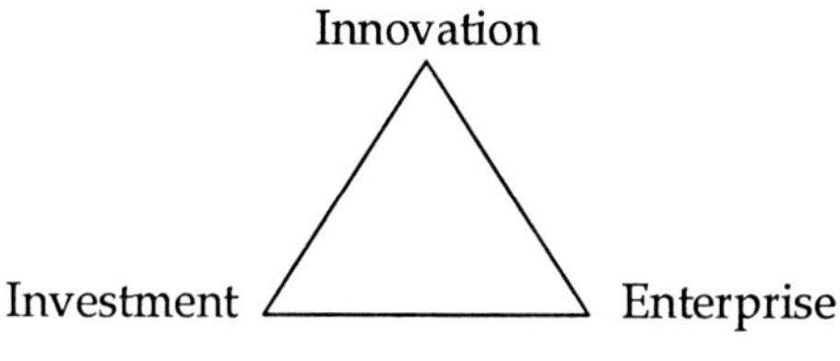

Fig. 4 : Golden Triangle

This model of golden triangle enables an entrepreneur sitting in New York to invest in an innovation from India to set up an enterprise in Tokyo. Or, regionally, an entrepreneur in Delhi can invest in an innovation from Ladakh to set up an enterprise in Mumbai.

4.3 Parameters for Technology Adoption in High Altitude

For any potential technology to be adopted it has to fit within the framework, otherwise even a good technology is not adopted by local community. Considering this fact research programme should start from the grassroots level itself (taking into their needs and priorities) in a bottom-up approach.

In a pilot study through PRA methodology it was found that more than the productivity, a crop should have good shelf life. It is necessary because of landlocked condition Ladakh remains cut off from rest of the world during winter for nearly six months in a year.

Therefore before any technology intervention, it is useful that efforts are made to understand the important parameters which are looked for in a technology by the local framers of Ladakh.

4.4 Local Knowledge System (LKS)

These days it has been recognized all over the world by the so called modern scientific community that there lies tremendous scope through the integration of LKS into the mainstream modern knowledge system, to evolve eco-friendly and sustainable appropriate technologies and practices suited to a particular niche or an ecosystem.

In this regard DIHAR can initiate a major programme to document, validate, and integrate LKS, so that cost effective appropriate technologies can be designed for Ladakh and similar other high altitude areas.

4.5 Grafting Institutions

'Grafting' institutions instead of 'crafting' new institutions may indeed be a viable approach to institution building and renewal (Gupta, 1991). This argument stems from a general understanding that at any time, or place where communities live, there can never be an institutional vacuum. Social construction without institutions is unthinkable. Given such an assumption, it pays to build upon existing bedrock of institutions to ensure deeper and organic fusion rather than crafting institutions afresh (as is often done in most watershed or joint forestry projects) as if there was no history of rules, norms, values or beliefs to guide future interactions.

In Ladakh, there are already various grassroots institutions already existing therefore they should be used for various extension activities. This way it helps in reducing the *ex-ante* cost (cost incurred before the agreement, they include variables like number of decision maker, heterogeneity of interest, skills assets of leader/experts, rules in use) and *ex-post* cost (like monitoring and enforcement cost; they include variables like size of resource, exclusion technology e.g. less in lagoons and more in offshore fisheries, appropriation technology, market arrangements, legitimacy of rule in use) and more importantly the degree of credibility is maintained which otherwise is difficult to achieve with a newly crafted institution.

"Emphasis should be on educating the farmers rather than just performing extension service (though it is also important). This will help farmers to decide of their own, what to adopt and what to discard out of their behaviour."

Through this paper, effort has been made to present a conceptual framework for extension activities in risk prone high altitude area of Ladakh and other similar regions. It is hoped that the framework will provide the basic outline that helps extension personnel and all those people involved in rural development to execute their work more effectively and appropriately.

References

Angchok, D. (2000). Traditional Method of Weather Forecasting in Tibetan Astrology and its Relevance in Today's Agriculture. Unpublished M.Sc. Thesis, Department of Extension, Education, Banaras Hindu University, Varanasi, India.

Angchok, D. (2003). A Framework to Study Local Institutions: Their Role in CPR Management in Risk Prone High Altitude Areas. Paper Presented In: National Seminar on *Extension Strategy for Promoting Development Initiatives Among Farming Communities*, ISEE, Division of Extension Education, IARI, New Delhi, 18-20 June 2003.

Fraser D. Transhimalaya unveiled (originally *The Marches of Hindustan,* ca. 1906), I (reprint), 1986, 106.

Gupta, A.K. 91997). Consumers can conserve: will you? *Honey Bee*, IIM, Ahmedabad, 8 (4): 1.

Gupta, A.K. (1991). Building Upon Peoples' Ecological Knowledge: Framework for Studying Culturally Embedded CPR Institutions. Paper Presented at the *Second Annual Conference of the International Association for the Study of Common Property*, Winnipeg, Manitoba, 26-29 September 1991.

Parasar, B. (1994). Indigenous Agricultural Knowledge of Tribals in Eastern Ghat High Land Zone of Orissa. Unpublished Ph.D. Thesis, Department of Extension, Education, Banaras Hindu University, Varanasi, India.

Reinfenberg, G. (1998). *Ladakhi Kitchen*, Melong Publication of Ladakh, Leh, 1998.

Rizvi J., Ladakh: Crossroads of High Asia, Oxford University Press, 1983.

Warren, D. Michael (1993). Indigenous Knowledge, Biodiversity Conservation and Development. Paper Presented in: International Conference on *Conservation of biodiversity in Africa: Local Initiatives and Institutional Roles*, Nairobi, Kenya, 30August - 3 September, 1993.

❏❏❏

Chapter-12

Biotechnological Approaches for Molecular Characterization and Conservation of Trans-Himalayan Flora

Janifer Raj X., Jitendra Kumar, Manish Bhoyar, Gyan P. Misra, Prabodh K. Bajpai, Ashish R. Warghat and R.B. Srivastava
Defence Institute of High Altitude Research (DIHAR), DRDO, Leh, Ladakh

1. Introduction

The Indian Trans-Himalayas span over 186,000 km^2 and a conservative estimate suggests that over 3000 dicot and 1000 monocot species are endemic to the Himalayas (Mani, 1978). The extraordinary high diversity appears to be a function of the extreme topographic variation and complexities comprising of deep valleys coupled with recently uplifted mountains. However, despite the richness, species diversity, this region is facing serious environmental threats due to destruction of forests and pasture, subsequent soil erosion, and anthropogenic activities. The potential instability and inherent vulnerability of the mountain ecosystems render the Trans-Himalayas as one of the ecologically fragile biogeographic zones of India. Opening up the previously inaccessible terrain due to road building and deteriorating pasture quality, threatens large number of plants (Fox *et al.*, 1994).

Globally, plant species are affected by human activities, either directly or indirectly and also change the level and pattern of genetic diversity in native plant species, leading to loss of genetic diversity that may ultimately reduce the evolutionary potential of a species to respond to environmental changes. Population fragmentation, isolation and lowered population densities may also modify patterns of gene exchange, reducing pollen and seed movement between sites and possibly increasing inbreeding within sites. For many species, genetic diversity is directly related to population size and levels of genetic diversity may increase isolation and decrease fitness leading to extinction.

Conservation of the genetic resources of endemic desert plants is crucial for the worldwide efforts, to prevent further degradation of the fragile ecosystems in arid and semi-arid regions and to sustain biodiversity. The desertification and ongoing ecosystem deterioration in arid and semi-arid regions worldwide has recently drawn attention of the international community on the urgent need to protect the environment of these regions. An understanding of how the diversity and composition of communities varies among locations is critical for prioritizing conservation efforts. For example, if the species composition is similar among sites across the region, then the specific location of a nature reserve is not important relative to this factor. However, if there are large differences in the flora among geographic locations, then the selection of areas for protecting native plant species poses a much greater challenge to conservation managers. Genetic diversity, species richness and assemblages are the three main levels for conservation of biodiversity. Levels of genetic variation within and among natural populations provide fundamental information on the evolutionary processes that bring about divergence within species and lead to speciation (Krauss, 1997). This information is crucial ultimately for the establishment of effective and efficient conservation practices (Hamrick and Godt, 1996).

2. Genetic Diversity Analysis through Molecular Markers

Genomes of higher plants are characterized by large sizes and complex organization. The use of molecular methods of DNA analysis is one of the approaches to study these complex genomes. Out crossing rates in plant populations have traditionally been estimated from polymorphism data at isozyme loci. In recent years, the use of molecular markers is preferred over conventional morphological and biochemical markers for genetic diversity studies. This is because molecular markers are not influenced by environmental effects and the developmental stages of the experimental material. Molecular DNA markers are polymorphic nucleotide sequences dispersed over the genome whose mutations may be detected by PCR-based techniques. POPGENE software is used to calculate Nei's unbiased genetic distance among different genotypes (Nei, 1978). GenAlex software is used to calculate principal coordinates analysis (PCA) that plots the relationship between distance matrix elements based on their first two principal coordinates (Peakall and Smouse, 2001). Hence, the introduction of molecular markers into biological studies offers new possibilities for studying genetic diversity and determining genetic relationships within and between species. Investigations on population genetic diversity and the structure of populations of the plants illustrate the evolutionary process and provide information useful for biological conservation. Currently, there are many molecular marker systems routinely used to evaluate genetic diversity in plants such as randomly amplified polymorphic DNA (RAPDs, Williams, 1990), amplified fragment length polymorphisms (AFLPs, Vos, 1995), Simple Sequence Repeats (SSR), (Zietkiewicz, 1994) and Inter Simple Sequence Repeats (ISSRs), etc.

Using these molecular tools Defence Institute of High Altitutde Research (DIHAR) has analysed the genetic diversity of Trans-Himalayan flora such as Rhodiola, Lucerne, Caper, Marsh Orchid, Mulberry, Apricot and Artemisia for designing appropriate conservation strategies.

2.1 *Prunus armeniaca* L. (Apricot)

Apricot (*Prunus armeniaca* L.) (Family: Rosaceae) is the most important fruit crop of cold arid regions of India which cover 3,200 ha with a total production of 5,200 MT (Dwivedi 2007). The phylogenetic relationships among 36 accessions of locally grown *Prunus armeniaca*, which were collected from two valleys viz. Nubra (9,600 ft) and Leh (11,500 ft) of Trans-Himalayan region were analyzed using 31 PCR markers (Kumar, 2009). This is the first report of molecular genetic diversity in apricot from this region of the world. The results indicate the presence of great genetic variability among local genotypes of apricot where between valley variations are more than within the valley variations. This could be exploited for the future conservation and breeding of apricot from this region. The protein profiling of 55 apricot cultivars using Lowry's Method revealed highest Protein content in Khalman genotype (2.25 mg/100g) and the least in Hangol ecotype (0.6 mg/100g).

2.2 *Rhodiola imbricata* (Golden Root)

Rhodiola imbricata Edgew (Family: Crassulaceae) is a promising rare medicinal plant found in Indian Himalayas producing various glycosides with adaptogenic properties. Salidroside, a phenolic glycoside of *R. imbricata* is the main bioactive agent and has been used by astronauts, divers and mountaineers to improve their stress resistance (Yan and Wu, 2004). The young leaves are used as vegetable by the Ladakhi people while in Amchi system of traditional medicine the roots are used against lung problems, cold, cough, fever, loss of energy and pulmonary complaints and used in preparation of health tonic which restores memory (Chaurasia *et al.*, 2007). ISSR-Polymerase Chain Reaction (PCR) has been successfully employed to reveal genetic variation among the wild populations of *R. imbricata* from the Trans-Himalayan region. For the 35 *R. imbricata* accessions tested with 38 ISSR primers, seven produced clear bands of which six primers were polymorphic. The resolving power of the 7 primers ranged from 8.74 for primer ISSR 4 to a maximum of 22.74 for primer ISSR 7. The genetic variation within Chang-La and Khardung-La populations of *R. imbricata* were high with Shannon's Information Index (I) values of 0.448 and 0.537 respectively and it may be credited to the clonal propagation through rhizomes and root slits as reported earlier in alpine plants with high levels of clonal diversity (Pluess and Stocklin, 2004).

The analysis of molecular variation (AMOVA) was performed and the variation was partitioned within populations (62%) and among populations (38%). Similar report was made by Lei *et al.* (2006) in *R.crenulata* using ISSR markers from Hengduan Mountains Region, in China. Principal coordinate analysis (PCA) clustered the populations in two major groups along a geographic gradient. The high levels of population differentiation indicated that isolation in *R.imbricata*, may be more severe or has occurred earlier in evolution. Even though the individuals from populations grouped together indicating that they breed within the same ecological zone, it was quite interesting to note a particular individual (K 11) which plotted away from the group of individuals from Khardung-La. The mountain barriers avoided gene flow between populations but there is an obvious indication of genetic drift within populations, where some extreme alleles do not breed with the existing alleles and are at the edge of genetic erosion. If those individuals are not preserved immediately they may be lost from the nature in course of time due to harvesting of rhizomes. High levels of natural fragmentation, strong genetic effects and isolation by distance are characteristics of Mountain ecosystems (Schonswetter *et al.* 2002). However, the fact that *Rhodiola* natural resources have decreased remarkably in recent years, owing to over-exploitation for medicine and shrinkage of their natural habitat, indicates that the threats to the survival of species of this genera mainly comes from human activity.

2.3 *Medicago species* (Alfalfa or Lucerne)

Ladakh harbors three species of *Medicago* such as *M.sativa, M. falcata and M.media* (Family: Leguminaceae) adapted to cold arid habitats and is the widely used fodder of the cold desert. The cultivated alfalfa (*Medicago sativa* L.) is autotetraploid ($2n = 4x = 32$), cross-pollinated and seed propagated (McCoy and Bingham, 1988). RAPD and ISSR markers are useful in evaluating the genetic diversity and structure of alfalfa, and other forage crops such as white clover (Gustine *et al.*, 2002), *Dactylis glomerata* (Tuna *et al.*, 2004), red clover (Grlju *et al.*, 2005), and perennial ryegrass (Bolari *et al.*, 2005). The major limitation of these markers is their dominant nature and this problem can be overcome by the use of efficient statistical methods such as analysis of molecular variance (AMOVA) (Excoffier *et al.*, 1992; Holsinger and Wallace, 2004). The present investigation with 20 random decamer primers and 11 ISSR primers produced 97.63 percent and 95.29 percent polymorphic bands and found high level of genetic diversity at molecular level among the *Medicago* accessions. Field observations also indicated that *Medicago* sp. is perennial and pollinated by bumble bees, which largely promotes out crossing. In addition, floral structure of this species facilitates the loading of pollen on the body of the pollinator by opening of the flower when the insect lands over the corolla. Both mechanisms help to maintain genetic polymorphisms across populations. The cross-pollination mechanism, allopolidy, sexual reproduction, high seed ratio and self sown ability to produce offspring of the *Medicago* species could have resulted in the accumulation of abundant genetic variation during the long evolution history.

In population genetics, the lower Gst values (RAPD) 0.0520, (ISSR) 0.0697 and (RAPD+ISSR) 0.0584 with higher Nm values (RAPD) 9.1181, (ISSR) 6.6773 and (RAPD+ISSR) 8.0682 suggested more pollen and seed dispersal among populations. High levels of gene flow from the above study indicated chance of adaptive gene flow from landraces to cultivars. Therefore, breeding approaches aiming to develop new synthetic cultivars have to collect a number of superior selected single plants with high levels of gene diversity. Studies on the biology of flowering and pollination indicate *Medicago* sp. is an out crosser. It was also supported by the genetic differentiation (G_{ST} = 0.0584) among populations of *Medicago* sp. which was very close to the average genetic differentiation (G_{ST} = 0.144) in out bred populations. The strong adaptability and wide distribution has resulted in little differentiation between populations. *Medicago* sp. exchanged its genes mainly by seeds and pollen dispersed by wind. Gene flow is the most important factor in making population genetic structure homogeneous and the large gene flow (Nm = 8.0682) of *Medicago* sp. could counteract most of the gene differentiation. AMOVA partitioned the genetic variation as 51 percent within populations and 49 percent among populations. The more or less same values of genetic variation within and among populations and high gene flow detected in this study point towards the fact that there was less instance of single isolated populations possessing unique genotypes not found in other populations. It is therefore imperative from conservation point of view to target few selected populations for conservation of genetic variation and breeding purposes. It is obvious from the variations scored that the alfalfa of Ladakh has undergone a very high level of natural hybridization and a number of characters have evolved which are quite different and more useful than the characters of *M. sativa* or *M. falcata*. A continuous monitoring of the species, including more populations, co-dominant genetic markers and the assessment of adaptive genetic variation will be useful to design effective conservation strategies and develop high yielding and cold resistant cultivars for different ecological niches of the cold desert.

2.4 *Capparis spinosa* L. (Caper)

Capparis spinosa L. (Family: Capparidaceae) locally known as '*Kabra*' is a well-known medicinal plant in 'Amchi system' which is occasionally used by local people of Ladakh as a leafy vegetable and forage. In India, it is found in inner valleys of Trans-Himalayan between 3020–3890 m which includes Indus, Nubra and Suru valleys of Ladakh region. It is an underutilized wild plant which grows in roadside, dry rocky slopes and stony soils, can withstand extreme temperature (-30°C to +35°C) of Ladakh and is highly drought tolerant. This plant has multiple uses in cuisine as salad, pickle and condiments. Bio-chemical studies have reported the presence of alkaloids, lipids, flavonoids and glucosinolates, cancer preventing agents and biopesticides in *Capparis spinosa*. Besides, it brings many environmental benefits, including soil and water conservation, desertification control and land reclamation in fragile cold ecosystem of Ladakh (Mishra *et al.*, 2009).

Limited reports are available on the molecular characterization of Capparis from any part of the world. Khouildi *et al.* (1999) have characterised cappers genetic diversity using RAPD marker, on Italian and Tunisian population. A genetic fingerprinting technique (AFLP) was studied by Inocencio *et al.* (2005) to determine the relationship among *Capparis* sp. from Spain, Morocco and Syriya. RAPD and ISSR marker techniques were used to determine the genetic variability in caper individuals collected from three valleys i.e. Indus, Nubra and Suru of Ladakh region in India. Ninety wild grown individual plants collected from nine populations were obtained from three valleys (Indus, Nubra and Suru) with altitude ranging from 10,286 ft (Nubra), 10,856 ft (Suru) to 11,268 ft (Indus). Twenty random decamer primers were used for RAPD amplification and eight for ISSR.

RAPD analysis yielded 223 fragments, of which 220 were polymorphic while, ISSR produced 85 bands, of which all were polymorphic, with an average of 11.0 and 10.62 polymorphic fragments per primer respectively. ISSR markers were found more efficient with regards to polymorphism detection, as H, I, Ht, Hs, Fst, diversity index and marker index were found more for ISSR than for RAPD marker. The pattern of clustering of the individuals remained more or less the same in RAPD and combined data of RAPD+ISSR. AMOVA analysis showed that total variation within the population was maximum, followed by among population and least for among valley in all the three cases. Pertaining to the management of Capparis, the high genetic differentiation of population indicates the requisite of conserving the utmost possible number of populations from different valleys of Trans-Himalayas.

2.5 *Dactylorhiza hatagirea* (Marsh Orchid)

Dactylorhiza hatagirea (D. Don) (Family: Orchidaceae) commonly known as Marsh Orchid is a critically endangered, high value medicinal plant species exclusively found in temperate regions of India with high market value of ₹ 2200-2500 per kg (dry tubers). The name *Dactylorhiza* is derived from Greek words "dactylos" (finger) and "rhiza" (root) referring to the palmately two to five lobed tubers (Chaurasia *et al.*, 2008). Thirteen populations from Trans-Himalayan region covering

Nubra, Suru and Indus valley were analysed morphometrically and genetically. A total of 28 quantitative and qualitative traits were measured on all 260 individuals for morphological analysis (Bateman and Denholm, 1985). PCA analysis of morphological data revealed 79 percent variation in among population. Genetic diversity was anlysed using 20 random decamer primers and 40 ISSR primers. According to results of RAPD AMOVA analysis, of the total genetic diversity, 57 percent was attributed to among populations and 43 percent within populations while, in case of ISSR AMOVA analysis, 60 percent was attributed to among population variation and 40 percent within populations. Low level of genetic diversity observed within populations of *Dactylorhiza* may be due to evolutionary factors such as mating system, gene flow, founder effect, random genetic drift and seed dispersal (Hamrick and Godt, 1996) and anthropogenic activities such as habitat fragmentation under human disturbance and habitat isolation. Orchids are generally self pollinated but in our study the level of gene flow (Nm=1.3947) per generation was also slightly higher than out-crossed animal pollinated species (Nm=1.154). From the level of gene flow, it is accomplished that *D.hatagirea* is an outcrossing species with deceptive pollination as reported earlier in *Dactylorhiza incarnata*. Moreover, *D. hatagirea* populations with more diversity should be conserved but separately because the mixing of populations may give rise to out breeding depression, loss of adaptation and breakup of co-adapted gene complexes.

2.6 *Morus* sp. (Mulberry)

Morus sp. (Mulberry) (Family: Moraceae) is a small to medium-sized monoecious or dioecious shrub or tree. It is locally called Oshe in Ladakh. It is recognized for its fruits rather than sericulture and people use Mulberry as fresh fruit in summers and dried ones in winters. In Ladakh *Morus alba* and *Morus indica* species were reported by Anand Rao (2005) but DIHAR has also found *Morus laevigata* which is a tree species. In our study, 21 inter-simple sequence repeats (ISSR) markers were used to investigate the genetic diversity of 25 mulberry accessions collected from three sampling sites viz. Turtuk village of Nubra valley (9240 ft) and Khalatse and Nurla villages of Leh valley (11,500 ft) of Trans-Himalayan region. The different plant accessions clustered according to their geographical location and the level of genetic diversity was found to be high within valleys and low among valleys which might be due to mountain barrier which restricted the gene flow among populations. The results recommend *in situ* conservation strategies to be followed for the mulberry germplasm conservation. Further studies on fruit quality such as antioxidant activity, total polyphenols and anthocyanin content are being carried out as Mulberry is used for edible purposes.

2.7 *Artemisia annua* L. (Worm Wood)

Artemisia annua L. (Family: Asteraceae) is an annual medicinal plant producing sesquiterpene compound called artemisinin. Artemisinin and its derivatives, in recent years have transformed the treatment of malaria and are effective in treatment against cancer. However, both wild and cultivated *A. annua* have very low artemisinin yield (<1.0%). Phylogenetic analysis of 20 *Artemisia annua* accessions and 25 accessions of five *Artemisia* species have been conducted with the help of different molecular marker like RAPD, ISRR, ILP and SSR. 20 RAPD, 17 ISSR, 35 Intran Length Polymorphism (ILP) and 16 SSR markers were used to analyse the genetic diversity among 20 *Artemisia annua* accessions. It was found that the genetic diversity among accesssions from Nubra valley were narrow than that of Leh valley, suggesting the importance and feasibility of introducing elite genotypes from different origins for *Artemisia* germplasm conservation and breeding programs. The genetic diversity analysis was carried out in five *Artemisia* species namely *A.annua, A.gmelinii, A.sieversiana, A.dracunculus* and *A.tourenefortiana* using 20 RAPD and 8 ISSR markers and the different origin of *A.dracunculus* could be attributed to the separate clustering pattern while all the other species belong to same origin. The number of alleles, mean heterozygosity and Polymorphism Information Content (PIC) values were also estimated. The *Artemisia* germplasm shows rich allelic diversity, which indicates that there is great potential for identification of associations and for developing varieties with traits of interest.

Due to ever-increasing resistance of malaria parasite strains to the available antimalarial drugs and growing concern of inadequate artemisinin supply to meet the future requirements for artemisinin based-combination therapies, it is imperative to search for supplementary artemisinin sources. We have reported *Artemisia tourenefortiana* that grows in the Trans Himalayas region of Ladakh as an alternative source of artemisinin albeit at low concentration than *A. annua*. Among the other species such as *A. dracunculus* and *A. gmelinii* artemisinin content was not detected, whereas in *A. sieversiana* only least quantity of artemisinin was detected. In case of

A. tourenefortiana artemisinin content was found to be maximum in flower (1.43%) followed by root (0.331%). Further the artemisinin content in flower varied from 0.05 to 1.43 percent with respect to different geographical location. Our results indicate a great potential for the use of *A. tourenefortiana* as an alternative source of artemisinin.

3. Conservation of Flora through Plant Tissue Culture

Considering the fact that the thick rose scented rhizomes are the economic part of *Rhodiola imbricata* the use of which depletes the plant from the natural habitat, experiments were carried out to evaluate the potential of *in vitro* propagation techniques for the conservation. An efficient micro propagation protocol based on multiple shoot induction and organogenesis was standardized. The shoot tip explants from *in vitro* germinated seeds and adult plant were cultured on MS medium supplemented with Kinetin (Kn), (0.5-1.0 mgL^{-1}) and in combination with Benzyl Adenine (BA), (0.5-2.0 mgL^{-1}). Maximum multiple shoot induction was observed on MS medium supplemented with BA (1.0 mgL^{-1}) and Kn (1.0 and 2.0 mgL^{-1}). Callusing was obtained on MS medium supplemented with BA, Indole Acetic Acid (IAA), Naphthalene Acetic Acid (NAA) and 2,4-Dichloro diphenoxy Acetic Acid (2,4-D). The callus sub-cultured on regeneration medium supplemented with BA (3.0 mgL^{-1}) and NAA (0.25 mgL^{-1}) responded for shoot organogenesis. The shoots obtained via multiple shoot induction and organogenesis were rooted on half-strength MS medium supplemented with Indole Butyric Acid (IBA), (1.0 mgL^{-1}). The protocol is expected to help in conservation of this rare medicinal plant. Further studies on this taxon are being carried out in our laboratory on phytochemical profiling for identification of novel compounds and standardization of agro techniques for field cultivation.

Orchids represent the most evolved and largest group of plants where the regeneration and multiplication through seeds in nature is limited due to suppressed endosperm and requirement of fungal stimulus (Harley, 1959; Sungkumlong and Deb, 2008). DIHAR has standardized the *in vitro* seed germination of *Dactylorhiza hatagirea* for the first time with seeds having minute or dust like particle. The green pods of *D. hatagirea* were surface sterilized by 70 percent ethanol for 2 minutes followed by 0.1 percent $HgCl_2$ for 7 minutes and then washed with double distilled water repeatedly 2-3 times. The green pods after sterilization was inoculated on three different basal media such as Murashige and Skoog, Norstog and Knudson with solid and liquid strength (full, ½ and ¼). The seeds germinated in ½ and ¼ solid and liquid media and the germinating seeds subsequently converted into Protocorm Like Bodies (PLBs) on the same medium. The PLBs later differentiated into shoot primordia with adventitious root. The *in vitro* mass multiplication of *D.hatagirea* will help in conservation and existence of this endangered species.

4. Search for Antifreeze Substances in Ladakh Flora

Antifreeze proteins (AFPs) are a group of proteins that protect organisms from deep freezing temperatures. Antifreeze activity has been widely distributed among organisms adapted to live at low temperatures, including prokaryotes, fungi, insects, plants and fish (Griffith *et al.*, 1995). The AFPs do not affect the melting point colligatively, but displace the freezing point of the ice to a lower temperature. This AFP-induced separation of the melting-and freezing temperatures is termed thermal hysteresis (TH) and the lower freezing temperature of the ice crystal is termed the Hysteresis Freezing Point (HFP). In plants antifreeze proteins are secreted in apoplast of leaves and crowns. In plants AFP have been isolated from *Solanum dulcamara* (Duman, 1993), *Secale creale* (Griffith *et al.*, 1992) *and Daucus carrota* (Smallwood *et al.*, 1999). The antifreeze activity in Seabuckthorn *(Hippophae rhamnoides)* and *Capparis spinosa* were tested at DIHAR as they are extremely cold hardy plants. In our preliminary study we have designed 8 antifreeze protein based PCR primers of three different origin (*Secale cereale, Pseudomonas putida* and *Daucas carrota*) and amplified with seabuckthorn and capparis genomic DNA. The amplified regions were sequenced in 34 accessions of Seabuckthorn and 29 accessions of Capparis and are submitted to NCBI database (Seabuckthorn: Accession No. GS927690-GS927723 and Capparis: Accession No. GS927724-GS927752).

5. Future Prospects

Mountain habitats experience special climatic conditions that often differ tremendously from the surrounding low lands and valleys. Steep topographic and therefore climatic gradients lead to heavily fragmented habitats, characterized by barriers to migration and genetic exchange. Levels of natural fragmentation are thus generally high and several studies have demonstrated strong genetic effects and isolation by distance (Schonswetter *et al.*, 2002). A successful management programme of endemic flora cannot be based only on molecular markers as markers might not be correlated to adaptive traits. A continuous monitoring of the species, including more populations, co-dominant genetic markers and the assessment of 'adaptive' genetic variation will be useful to design effective conservation strategies. The individuals noted as rare ones from these studies should be preserved as vegetative propagules and seeds to conserve and use in further breeding programmes and reintroduction.

References

Anand Rao, A., Thangavelu, K. and Sharma, K.R. (2005). Distribution and variation of Mulberry genetic resources (Morus spp) in high altitude cold deserts of Ladakh Himalayas region. *Indian Journal of Plant Genetic Resources*.18 (2): 255-259.

Bateman, R.M. and Denholm, I. (1985) A reappraisal of the British and Irish dactylochids, 2. The diploid marsh-orchids. *Watsonia*. 15: 321-355.

Bolari, S., Barth, S. Melchinger, A.E. and Posselt, K. (2005).Genetic diversity in European perennial ryegrass cultivars investigated with RAPD markers. *Plant Breeding*. 124: 161-166.

Chaurasia, O,P., Ahmed, Z. and Ballabh, B.(2007). *Ethnobotany and Plants of Trans-Himalayas*. Satish Serial Publishing House, Delhi. India.

Chaurasia, O.P., Ballabh, B., Janifer, R., Saurav, K., Diwakar, S. and Amol, T. (2008). In: Advances in Agriculture, Environment and Health.(eds. Chaurasia, O.P., Yadav, A., Rimando, A.M. and Terrill, T.H.) Satish Serial Publishing House, Delhi. pp. 512

Duman, J.G., Wu, D.W., Olsen, T.M., Urrutia, M. and Tursman, D. (1993). Thermal-ysteresis proteins. *Advances in. Low-Temperature Biology*. 2: 131–182.

Dwivedi, S.K., Kareem, A. and Ahmed, Z. (2007). Apricot in Ladakh. Field Research Laboratory, DRDO, C/o 56 APO, Leh, Ladakh.

Excoffler, L., Source, P.E. and Quative, J.M. (1992). Analysis of molecular variance inferred from metvic distances among DNA haplotypes : applications to human mitochondria DNA restviction sites heatics, 131: 479-491.

Fox, J.L., Nurbu, C., Bhatt, S. and Chandola, A. (1994) Wildlife conservation and landuse changes in the Trans himalayas region of Ladakh. *Mountain Research and Development*. 14: 39-60.

Gustine, D.L., Voigt, P.W, Brummer, E.C. and Papadopoulos, Y.A.(2002).Genetic variation of RAPD markers for North American white clover collections and cultivars. *Crop Science*. 42: 343-347.

Griffith, M., Ala, P, Yang, D.S.C, Hon,W. and Moatt B.A.(1992). Antifreeze Protein Produced Endogenously in Winter Rye Leaves. *Plant Physiology*.100: 593-596.

Griffith, M. and Ewart. K.V. (1995). Antifreeze proteins and their potential uses in frozen foods. *Biotechnology Advances.* 13: 375-402.

Grlju, S., Bolari, S., Popovi, S., Upi, T., Tucak, M. and Kozumplik, V. (2005). Assessment of morphological and RAPD variation among and within red clover cultivars after natural selection. *Die Bodenkultur.* 56: 183-188.

Harley, J. L. (1959). *The Biology of Mycorrhiza,* Leonard Hill, London.

Hamrick, J.L. and Godt M.J.W. (1996). Conservation genetics of endemic plant species. In: Avise J.C. and Hamrick J.L. (eds) Conservation Genetics: Case Histories from Nature. Chapman & Hall, New York, pp. 281–304.

Holsinger, K.E. and Wallace, L.E. (2004). Boyesian approaches for the analysis of population genetic structure: an example from *Plantanthera leucophaea* (Orchidaceae). Molecular Ecology. 13: 887-894.

Inocencio, C., Cowan, R.S., Alcaraz, F., Rivera, D. and Fay, M.F.(2005). AFLP fingerprinting in Capparis subgenus Capparis related to the commercial sources of capers. *Genetic Resources and Crop Evolution.* 52: 137-144.

Khouildi, S., Pagnotta, M.A., Tanzarella, O.A., Porceddu, E. and Ghorbel, A.(1999). Assessement of the genetic variation in natural populations of *Capparis spinosa* L. using RAPD analysis. *CWANA Newsletter.* p 8.

Kumar, M., Mishra, G.P., Singh, R., Kumar, J., Naik, P.K. and Singh, S.B.(2009). Correspondence of ISSR and RAPD markers for comparative analysis of genetic diversity among different apricot genotypes from cold arid desert of trans Himalayas. *Physiology and Molecular Biology of Plants.*15(3) : 225-236.

Krauss S.L. (1997). Low genetic diversity in Persoonia mollis (Proteaceae), a fire-sensitive shrub occurring in a fire-prone habitat. *Heredity* 78: 41–49.

Lei, H.G., Tashi T., Suhua S. and Yang Z. (2006). Determination of genetic variation in *Rhodiola crenulata* from the Hengduan Mountains Region, China using intersimple sequence repeats Yidong. *Genetics and Molecular Biology,* 29(2): 339-344.

Mani M.S. (1978). Ecology and phytogeography of high-altitude plants of Northwest Himalayas. Oxford and IBH, New Delhi, India.

McCoy, T.J. and Bingham, E.T. (1988). Cytology and cytogenetics of alfalfa. In *Alfalfa and alfalfa improvement.* Agronomy Monograph 29 (eds. A. A. Hanson, D. K. Barnes and R. R. Hill),pp. 737–776. Madison, USA.

Mishra, G.P., Singh, R., Bhoyar, M., Singh, S.B., (2009). *Capparis spinosa*: unconventional potential food source in cold arid deserts of Ladakh. *Current Science.* 96: 1563–1564.

Nei, M. (1978). Estimation of average heterozygosity and genetic distance from a small number of individual. *Genetics* 89: 583–590.

Peakall, R. and Smouse P.E. (2001). GenAlEx V5: Genetic Analysis in Excel. Population genetic software for teaching and research. Australian National University, Canberra, Australia, (http://www.anu.edu.au/BoZo/GenAlEx/).

Pluess, A.R. and Stocklin, J.(2004).Population genetic diversity of the clonal plant *Geum reptans* (Rosaceae) in the Swiss Alps. *American Journal of Botany* 91: 2013–2021.

Schonswetter, P., Tribsch, A., Barfuss, M. and Niklfeld, H.(2002). Several Pleistocene refugia detected in the high alpine plant *Phyteuma globulariifolium* Sternb. & Hoppe (Campanulaceae) in the European Alps. *Molecular Ecology*. 11: 2637–2647.

Smallwood, M., Worrall, D. and Byass, L. (1999). Isolation and characterization of a novel antifreeze protein from carrot (*Daucus carota*). *Biochemical Journal.* 340 (2): 385-391.

Sungkumlong, and Deb, C. R.(2008). Effects of different factors on immature embryo culture, PLBs differentiation and rapid mass multiplication of *Coelogyne suaveolens* (Lindl.) Hook. Indian *Journal of Experimental Biology.* 46: 243-248.

Tuna, M., Khadka, D.K, Arumuganathan, K. and Golan, G.A. (2004). Characterization of natural orchardgrass (*Dactylis glomerata* L.) populations of the Thrace Region of Turkey based on ploidy and DNA polymorphisms. *Euphytica.* 135: 39-46.

Williams, K., Kubelik, A., Livak, K., Rafalski, J. and Tingey, V. (1990) DNA polymorphisms amplified by arbitrary primers are useful as genetic markers. *Nucleic Acid Research.* 18: 6531-6535.

Yan, X.F. and Wu, Y.M. (2004). Soil nutrient factors related to salidroside production of *Rhodiola sachalinensis* distributed in Chang Bai. *Mountain and Environment Experimental Botany.* 52: 267-276.

Zietkiewicz, E., Rafalski, A. and Labuda, D. (1994). Genome fingerprinting by simple sequence repeat (SSR)-anchored polymerase chain reaction amplification. *Genomics.* 20: 176–183.

□□□

Chapter-13

Energy Efficient Long-Term Conservation of Plant Germplasm

Narendra Singh, R.B. Srivastava and S.B. Maurya
Defence Institute of High Altitude Research (DIHAR), DRDO, Leh, Ladakh

Depleting natural bio-resources and its impact on national and international peace and security has emerged as the focus in the present world. Globally, plant species are affected by human activities, either directly or indirectly and also through change in the level and pattern of genetic diversity in native plant species, leading to loss of genetic diversity that may ultimately reduce the evolutionary potential of a species to respond to environmental changes. Population fragmentation, isolation and diminished population densities may also modify patterns of gene exchange, reducing pollen and seed movement between sites and possibly increasing inbreeding within sites. For many species, genetic diversity is directly related to population size. The levels of genetic diversity may increase isolation and decrease fitness leading to extinction.

According to the United Nations Food and Agriculture Organization, as much as 75 percent of the genetic diversity (hundreds of thousands of plant varieties) have been lost in the lost century. Hence, the genetic diversity generated over thousands of years in plant species need long-term conservation to ensure food security for human beings. The existing plant genetic resources (both cultivated species and their wild relatives) are the biological foundation of agriculture and serve as raw material for all plant breeding programmes. International standards promulgated in 1994 for the conservation of plant genetic resources for food and agriculture call for each "unique" accession to be duplicated and stored in at least one additional location, ideally in a different country or at different locations within the country. Food and Agriculture Organisations (FAO's) State of the World's Plant Genetic Resources documented that individual genebanks are vulnerable to a host of problems that can endanger their collection that include poor management, lack of adequate funding, equipment failures and natural catastrophes. In addition to these threats, it has also been recognized that there is a need to provide protection against potentially more cataclysmic dangers civil strife, war and acts of terrorism. The need has compelled to ensure the long-term conservation of plant genetic resources, protecting them from both old and new threats, as noted in both routine as well as in unprecedented occurrences.

The genetic diversity contained within cultivated plant species is immense. Over 6 million accessions - samples - of this diversity are conserved (principally in the form of seed) in cold storage facilities in various locations around the world. It is stated that these resources stand between us and catastrophic starvation on a scale we cannot imagine. Our existence on earth rests on how well we take care of these seeds and their existence depends on us. It is estimated that over 3000 dicot and 1000 monocot species are endemic to the Himalayas. Subjecting the previously inaccessible terrain due to road building and deteriorating pasture quality, threatens large number of plants in Himalayan region (Fox *et al.*, 1994).

Levels of genetic variation within and among natural populations provide fundamental information on the evolutionary processes that bring about divergence within species and lead to speciation (Krauss, 1997) and establishment of effective and efficient conservation practices (Hamrick and Godt, 1996).

Loss of viability during long term seed storage (more than 30 years) in ambient conditions is a serious problem. There is a requirement of low temperature based on the duration of storage of seeds. To conserve the diversity, the seeds of different types of plants need long-term storage at low temperature. According to the study conducted by Pritchard and Dickie (2003) seed storage at -18°C to -20°C temperatures are found ideal for storage of seeds for maximum duration (Table 1).

Table 1 : Predicted Longevity of Seeds of Selected Crops at 5 percent Moisture Content Stored at -20°C

Crop	Expected Longevity (Years)
Barley	2061
Lettuce	73
Maize	1125
Onion	413
Pea	9876
Rice	1138
Wheat	1693

As depicted in the table above, seeds stored at -18°C to -20°C remain viable for many years (depending on the species) or being dried to the proper moisture content and packaged properly, it is estimated that seeds of most major food crops can be successfully stored at -18°C for hundreds of years and seed of many crops including some of the major grains can be stored for thousands of years. The Nordic Gene Bank has found no meaningful decline in viability of seeds stored in permafrost conditions at Svalbard in the 18th year of a 100-year experiment.

To create low temperature conditions especially in predominantly hot, humid tropical and sub-tropical climate, there is a requirement of very high recurring expenditure on energy for thousands years. Hence for cost effective and safe conservation of important plant genetic resources (PGR) as safety duplicates, it requires exploration and use of alternative methods of long-term germplasm conservation. On the other hand, the breeders use variety of germplasm to develop desired variety or parental lines for developing hybrids of particular trait, in this process, large number of germplasm is discarded which may have extinct quality for other traits and in absence of adequate long-term conservation facility this valuable germplasm is depleted. Hence it should be saved for future use as research food safety. Energy efficient genebank in permafrost conditions can play an important role for cost effective long-term conservation of diversity in plants.

Norway has created permafrost based long-term seed storage facility at Svalbard to take the advantage of naturally low temperature conditions at the site. During experimentations, Defence Institute of High Altitude Research (DIHAR) has found that Changla at an altitude of 5360 m AMSL situated about 90 km from Leh city also has low temperature conditions like Svalbard, Norway.

A joint exploratory experiment on conservation of PGR was launched in 2001 (Defence Institute of High Altitude Research, Leh-Ladakh and National Bureau of Plant Genetic Resources, New Delhi) for long-term storage of seed at Khardungla (5500 AMSL). Seed sample of 149 representative accessions of 47 major field and horticultural crops have been stored. Seed quality was examined after 1st, 2nd & 4th year of storage. It was found in the study that most crop species are able to maintain the seed quality near to their initial value. Variation in maintaining seed quality between the accessions of the same crop species, difference between crops of a group and between the crop groups as maintained in seed germination were not significant unlike those of between accessions and the results were similar to that of seeds stored under artificial low temperature conditions of -20°C in refrigerated chambers (Singh *et al.*, 2008). In this preliminary research trial, low temperature at 5500 m altitude in cold desert climatic conditions of Ladakh was found suitable for energy efficient cost effective long-term conservation of PGR.

DIHAR has created an energy efficient prototype facility for long-term plant germplasm conservation in permafrost conditions of Chang-La (5360 m above MSL). On the basis of requirement of storage of large number of accessions of plant germplasm this facility may expand in future. The fundamental premise which this feasibility study is based is that a facility at Changla, Ladakh should offer the highest feasible level of long-term protection to the national community for nation's unique plant genetic resources and provide food and agriculture security against "worst case scenario" including nuclear war, civil strife, catastrophic failures or major acts of terrorism affecting existing genebanks *etc.* It may function as a source of ultimate safety net, a backup reservoir to replenish materials that might have been permanently lost in all other collections. In order to meet international needs, address technical and political concerns and to ensure a high level of confidence, the expert scientists believe that a facility at Changla, Ladakh can meet the purpose. It may be designed further so as to offer effective, long-term conservation meeting the highest international technical standards for conservation of valuable plant genetic resources.

Genebank at Changla, Ladakh is the ideal place for long-term plant germplasm preservation at low energy input and cost effective unrecurring expenditure and will help to check the erosion of important genes by preserving them at naturally available low temperatures. The preserved diversity of the genes will serve as a safety net for current and future food security in an era of global warming and climate change. This facility will provide an insurance against both incremental and catastrophic loss of the plant diversity. The artificial cooling based techniques for germplasm storage are very expensive, so permafrost based germplasm storage is a preferred alternative. After expansion, the facility might be able to protect Nation's most valuable plant genetic resources and it may do so in a manner that is efficient, sustainable, inexpensive, and politically and legally acceptable. Thus making a major contribution to food and environmental security and to the safety and well-being of our fast growing population repositary may be offered for germplasm storage at international level.

Due to climate and geology, an underground facility at Changla, Ladakh can provide the advantage of permafrost conditions and utilize the considerably colder winter temperatures (Table 2) to provide storage conditions that may meet the highest international standards for long-term conservation at a low cost and with relatively little dependence on mechanical technologies.

Table 2 : Details of temperature in different months at Changla inside the germplasm store

S.No.	Month	Maximum Temperature (°C)	Minimum Temperature (°C)	Average Temperature (°C)
1	December2010	-15.8	-18.1	-17.2
2	January2011	-13.7	-24.4	-19.2
3	February2011	-14.7	-17.3	-16.2
4	March2011	-12.1	-14.9	-13.7
5	April2011	-8.8	-12.1	-10.7
	Average	-13.02	-17.36	**-15.4**

The site is located at 5360 m above MSL in remote region and cold desert conditions. A suitably constructed facility is expected to be more secure and permanent than any other existing genebanks.

References

Fox, J.L., Nurbu, C., Bhatt, S. and Chandola, A. (1994) Wildlife conservation and landuse changes in the Trans-Himalayas region of Ladakh. *Mountain Research and Development*. 14: 39-60.

Hamrick, J.L. and Godt M.J.W. 1996. Conservation genetics of endemic plant species. In: Avise J.C. and Hamrick J.L. (eds) Conservation Genetics: Case Histories from Nature. Chapman & Hall, New York, pp. 281–304.

Krauss S.L. 1997. Low genetic diversity in *Persoonia mollis* (Proteaceae), a fire-sensitive shrub occurring in a fire-prone habitat. *Heredity* 78: 41–49.

Pritchard, H.W. and Dickie, J.B. (2003). Predicting Seed Longevity: The use and abuse of seed viability equations. In Smith, Dickie, et al., Seed Conservation: turning science into practice. Royal Botanic Gardens, Kew, UK.

Singh, A., S.K. Jain, K. Srinivasan, M. Uprety , Narendra Singh, B. Raut, D. P. Attrey and Rakesh Singh (2008): Assessing natural low temperature conditions in Himalayas region for long-term storage of seeds: Facilitating cost-effective conservation of plant genetic resources. Seed Research, vol. 36 (2): 191-203.

□□□

Chapter-14

Modern Dairy Farming in Cold Arid Regions

Vijay K. Bharti, A. Biswas, G. Venkatesan, D. Gogoi, P. B. Deshmukh, S. Patial, P.R. Venkatesh and R.B. Srivastava
Defence Institute of High Altitude Research (DIHAR), DRDO, Leh, Ladakh

1. Introduction

Dairy farming is an important component of the livestock sub-sector in India and indeed of the national economy, especially with the increasing population that is expected to outstrip production in coming years. Cattle rearing (*Bos taurus*) in Leh-Ladakh is the most important livestock activity of rural population. However, the adoption rate of modern dairy farming practices is very low in Leh-Ladakh due to adverse climatic conditions. Ladakh situated in the northern frontiers of India, is one of the world's highest inhabited lands (ICAR, 2002). In this region the temperature fluctuates from + 35°C in summers (day temperature) to -40°C in winters which exerts huge stress on animals. The atmospheric oxygen pressure is 30 percent lower than at MSL and this exposes hypobaric-hypoxic condition to animals. The cold desert conditions are not suited for the natural habitat of animal husbandry (Sherpa, 1991).

2. Salient Features of Dairy Farming in Cold Desert of Leh-Ladakh

In Ladakh region, villagers live in settled communities distributed in pockets. Livestock is stall fed during long winter months, but is left to pastoral grazing in summer. Dairying is one of most important dimensions of diversified agriculture. This is thought to be a more attractive economic activity for the small farmers after poultry rearing in this region. Women are mainly involved in day-to-day farm activities.

2.1. Animal Genetic Resources

Cattle population (information rendered by Chief Anim. Husb. Officer, Leh) in Leh district (2010) are as follows : Local breed-24813 (Male-1218, Female-23595), Jersey Cross- 11418 (Male- 2889, Female-8529). Leh-Ladakh being a temperate region, the local breed of cattle is humpless and resembles the exotic cattle. These animals are well adapted and have less mortality due to high altitude stress but have very low production potential.

2.2. Animal Breeding Practices

Breeding practices prevailing in the area vary from the plain areas of the country. Breeding of animals is mainly through natural service by available bulls as adequate facilities are not available for artificial insemination (A.I.). Prolonged age at first calving (4-6 years), repeat breeding, long calving interval and abortion are very common. Many farmers use A.I., but due to poor conception rate of A.I., they again go to natural service by bulls reared for the draught/agricultural work due to non availability of good breeding bulls.

2.3. Animal Housing and Management Practices

Housing is challenging for both human as well as animals due to severe adverse climatic conditions. Therefore, farmers build animal house of mud wall, roof covered with dry grasses, poor ventilated room so that house inside remain warm. Most of the animal sheds lack basic facilities like ventilation, drainage system and ligthing in temperate conditions of Ladakh. In abesence of proper ventilation animals become susceptible to respiratory tract infection and mucosal irritation due to exposure to high ammonia concentration. They clean the sheds once in a week/month for dung/urine in winter. The management practices followed are very old. During the snowfall and extreme winter season farmers provide fumigation and artificial heating (Bukhari) to the animals. Unlike in organized dairy farms, farmers are not practicing weaning system. Majority of the farmers do not know the proper time of the vaccination. Castration of male calf is done with a very crude method which is very painful. Surprisingly, majority of the farmers do not pay attention to care and management of heifers that leads to higher age of first calving. Farmers stop pasture grazing of advance pregnant animals because they know that steep slopes are very dangerous to the animals.

Fig. 1 : Stall feeding of dairy cattle at village level

2.4. Animal Feeding

Non availability of nutritious animal feed resources round the year is the major constrain that hampers efforts towards dairy development in this region. Hence, marked shortage for quality animal feed resources is very common in this region. Farmers are totally dependent on the locally available unconventional feed and fodder like willow, poplar and other tree leaves and wild grasses grown in the barren land for feeding to their animals during summer months (May-October). The major fodder crops of the region are alfalfa, oat, pea, etc. Fodder is very costly in Ladakh fetching as high as ₹ 11.00-13.00/ kg dry fodder. Therefore, to maintain the animals on stall fed is a very costly affair. Very few farmers are able to arrange wheat straw for crossbred cattle and only some farmers offer salt and mineral mixture to their animals. Concentrate mixture, made at home is being given only to lactating animals but not to unproductive animals. Further, majority of the farmers are dependent on natural resources for their animal husbandry because of lack of irrigation facilities and due to topography of the region. Therefore, search for low cost nonconventional feed ingredients to support animal nutrition without any adverse effects is of prime importance (Fig. 1&2).

Fig. 2 : Barren lands covered with wild grasses during summer months

2.5. Milk Production

Local animals are poor milk yielders i.e. 1-2 L/day where as HF/ Jersey cross breeds produce upto 2-6 L/day. Cross bred animals are yielding on an average 1500 L milk against 450 L by local cow per lactation. Therefore, farmers start rearing crossbred cows for high milk production. They sell the surplus milk to Indian Army through milk co-operative societies, Leh which is supported by the government since 2001. There are 05 milk co-operative societies at present. On an average 9.73 lakhs litre is marketed annually. Gradually many villages are coming under this umbrella. Scarcity of feed and fodder, lack of knowledge about preservation, lack of road and other transport facility etc. seem to be the major constraints for economic milk production. Many farmers are not aware about possible milk contamination with human and animal origin pathogens through their unhygienic milking practices. At the time of milking, the hands of the milker as well as the udder of the animal and the milking utensils are seldom properly washed.

2.6. Animal Health Managements

Anoestrous and still birth are major problems in cattle which may be due to cold stress, hypobaric hypoxia, poor nutrients supply and minerals deficiency. The animals are underfed that results in their late maturity, high mortality, poor lifetime performance and infertility. Poor semen quality due to high altitude (HA) stress is another problem with elite bulls of this region. Ectoparasite infestation and gastrointestinal helminths (liver fluke, ascariasis, and strongaloids) are also common despite severe dry cold climates. Brucellosis, T.B., Jones disease, FMB, Mastitis etc. are the major pathogenic origin diseases. Cataract (due to intense UV ray exposure), brisket disease, congestive heart failure, hydropericardium, pulmonary oedema, ascites, urinary infection etc. are the frequent pathophysiological symptoms encountered during animal health practices. Farmers, especially women confidently use local remedies comprising herbal medicines. Therefore, development and introduction of scientific animal health package and practices are necessary for farmers.

2.7. Extension / Institutional Supports

The extension machinery has to be an ideal bridge between the research/development institutions and the beneficiaries. At the moment, in spite of KVKs and other local body's supports, lack of efficient extension mechanism towards livestock farming is a big constraint coming in the way of commercial dairy farming especially at high altitude. Therefore, an effective extension setup, well equipped with know-how and solid infrastructural back up is required to cater the local needs of the farmers of high altitude (Farooquee *et al.*, 2004).

3. Government Initiatives for Dairy Development

Animal Husbandry Department (J&K Govt.) is also working for breed improvement and quality fodder production in this area. In this direction, they established one pure Jersey farm at Agling (Bull Mother Farm) and another Jersey cross breed farm at Chanspa for further breeding and improvement of acclimatized and pedigree pure Jersey Bulls and cross bred, respectively. These animals perform well at farmer's level. Further they established 01 District Veterinary Hospital, 01 Disease Investigation Laboratory, 50 Livestock Development Centres, 07 Intensive Cattle Development Centres, 03 Animal Husbandry Sub Units to provide breeding (by natural service and artificial insemination) and health cover facilities to the livestock.

They started Rehbar Pashu Palan with the idea to provide veterinary health services in remote villages. Under this scheme, unemployed educated youth of the concerned village are educated and trained for minimum essential veterinary care.

In spite of these efforts, dairy husbandry has not become popular in this region. However, DIHAR has taken many initiatives through R&D to help local farmers for adoption of modern dairy farming, so that fresh milk availability can be enhanced to meet Indian troop's requirements.

4. DIHAR's Efforts Towards Advancements of Dairy Farming in Leh-Ladakh

Defence Institute of High Altitude Research (DIHAR) has developed several technologies aimed at improving milk production and quality in Leh-Ladakh. Animal husbandry to a great extent remained a subsidiary of agriculture in the competition for land to grow the food in this region, also poor animal managemental practices, limited veterinary medical care, fodder scarcity and difficulties in housing were other reasons.

In 1963, a dairy farm of 35 Kashmiri cattle from J&K government was taken over by DIHAR, which existed since 1955 with not much success. However, DIHAR's animal husbandry activities with support of local animal husbandry departments were the first efforts towards scientific dairy cattle rearing in this region. Their main activities were to improve local cattle by up-gradation with elite Kashmiri bulls.

However, limited improvement was seen in spite of increased genetic potential due to meagre amount of quality fodder availability to these animals. Having considered these factors, DIHAR has decided up on the strategy of improving the genetic make up of local cattle by cross breeding, augmenting the local availability of feed and fodder and evolving a scientifically and economically package of practices for managing the cattle herd. Initially, the thrust had been to study the performance of animals inducted from plains, cross breeding and up-gradation for improvement of genetic potential of the local animals, housing of animals under extreme cold conditions, and problems related with fertility, sterility and different disease conditions, etc. Since then DIHAR has been engaged in developing a suitable milk breed for cattle and preparing low cost balanced feed using locally available ingredients.

Fig. 3 : HF cross calves and bull

Jersey was the first exotic breed inducted and studied for its adaptation and performance during 1970s under a project titled 'Genetic improvement for milk production in local livestock through artificial insemination (AI). Thereafter, DIHAR undertook the studies on performance of Frieswal cows and Sahiwal under different agro climatic conditions of Ladakh during 1978-85. In 1980, these bulls were trained for semen donation. Semen dilution and preservation techniques were studied for establishing Artificial Insemination facility at DIHAR. It became a unique facility at DIHAR and among cold desert part of world. Inseminations using the diluted semen gave encouraging results. Performance and survivability of these cross bred (HF x Sahiwal) cattle was found better than other crossbred cows born and bred at this region. It was found that 5/8 blood in crossbred (HF x Sahiwal) cattle has been the best in all parameters of production, growth and adaptability. Farmers were educated on the significance of proper calf management and care of crossbred progeny (Fig. 3 & 4).

Fig. 4 : Sahiwal at DIHAR's Dairy farm

Crossbred animals which were inducted in 1981-85 performed well at different altitude conditions but few animals suffered from brisket diseases. The performance of few animals at higher altitude was found deteriorating due to prolonged cold stress, hypobaric hypoxia and less availability of green fodder as compared with animals at lower altitude. Acclimatization to cold stress and tolerance to hypobaric hypoxia was observed better in higher crosses as compared to lower crosses. A suitable ration for cattle was formulated during the year 1988. The significance of alterations in the prolactin levels was also observed.

Effect of acute and chronic exposure to high altitude on pulmonary haemodynamics of cattle was studied during 1988-89 in collaboration with PGI, Chandigarh. During the study hyper responders were identified. Research related to this showed that higher arterial pulmonary pressure is responsible for the heart failure. Subsequently sixty crossbred HF cows were inducted from Military Farm, Meerut for studying their adaptation and performance in Nubra and Suru valley.

The performance of buffaloes under high altitude conditions with special reference to physiological acclimatization, feed consumption, and milk production also were studied during 1979-85 using 05 buffalo heifers and 01 buffalo bull. It was found that buffaloes could not withstand extreme cold and dry climate of this area.

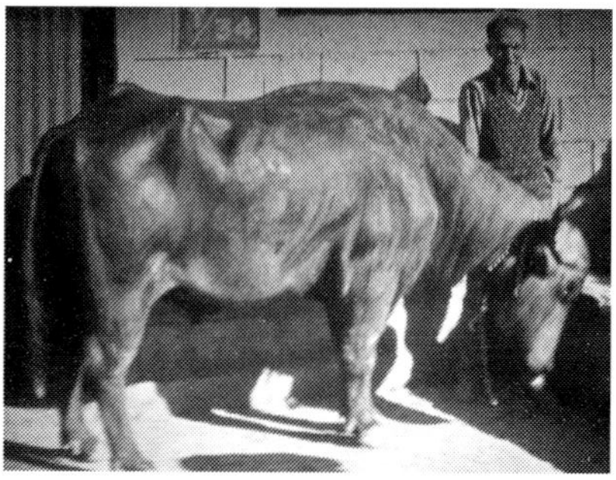

Fig. 5 : Buffalo at DIHAR's Dairy farm

Continuing its efforts on standardising dairy husbandry practices and integrating it with the locally popular yak and its crosses, DIHAR had procured few pure Ladakhi Yak and Demos (cross bred Yak) during the year 1995. Interestingly, faecal examination of these animals showed that yaks of this region are free from gastrointestinal parasitic infestation (Dong *et al.,* 2009).

Increased incidence of still birth was observed when temperature was below -15°C. However, DIHAR was able to standardise dairy management practices including feeding, housing, calf rearing and breeding.

DIHAR also had made some pioneering attempts on a novel assisted reproductive technique like embryo transfer technology (ETT) in local and cross bred animals with variable success in 1996. In spite of successful super ovulation and collection of embryos, the pregnancy couldn't be achieved through ETT because of unfavourable weather conditions and extreme low temperature.

A positive fall out of these studies was the successful demonstration of estrous induction using low doses of PGF2α through intra vaginal sub mucosal route.

DIHAR had been distributing pedigreed male claves free of cost and female calves at subsidised prices to the progressive farmers. Though the impact of this programme is not distinct, feedback of the same was encouraging. From 2001, the laboratory has maintained its experimental dairy herd on the concept of revolving fund and achieved self sustenance. Dairy unit of this laboratory could achieve an average milk yield at 13.4 L recently with a calving interval of 14 months.

This clearly demonstrates that scientifically run dairying ventures in Ladakh can be economically viable. DIHAR has continuously been supplying fresh milk to Indian troops from Partapur (Nubra Valley) and Leh dairy Units. DIHAR conducts regular training programme for Jawans from Indian Army and local farmers on scientific methods of commercial dairy farming and livestock managements.

DIHAR has embarked on an ambitious project on fodder production and an evolving a suitable milk breed by trying various crosses from HF, Jersey, Frieswal and local cattle and yaks. Natural service to local cow with our elite bulls (at Leh, Ranbirpura, Partapur) and AI were the constant supports from DIHAR to local farmers to upgrade their low productivity animals. These efforts has lead to increased milk supply to army and local population (Fig. 6 & 7).

Fig. 6 : DIHAR's Dairy unit at Leh

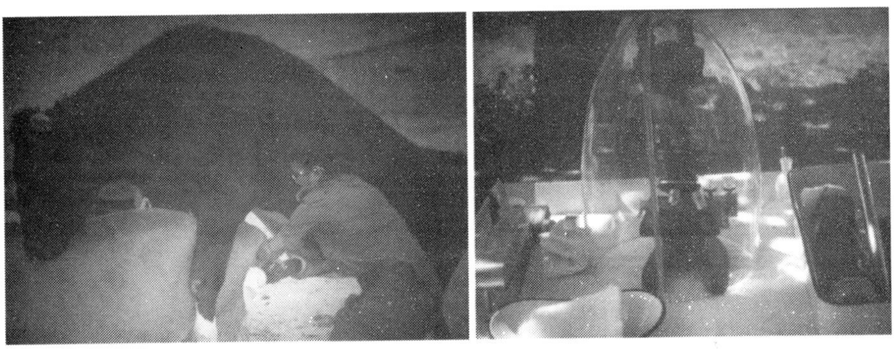

Fig. 7 : Semen collection and evaluation at DIHAR

Installation of a pasteurization plant at DIHAR in 2004 is a part of larger perspective plan for user towards white revolution in all respects. Currently, we are working for development of ration scale based on nutrient requirement at high altitude conditions, animal diseases prevalence and adaptation and performance of native and crossbred cattle in Ladakh.

5. Summary

Deficiency of quality feed and fodder, poor conception rate, prolonged age at first calving (4-6 years), unhygienic housing, poor disease management system, little knowledge about vaccination of livestock and high morbidity and mortality in livestock have been identified as the major constraints in animal husbandry sector in this area. However, local government's efforts along with DIHAR's R&D activities towards fodder production, breed improvement and livestock health managements may bring new dimensions to dairy sector in cold desert regions. Hence, it is perceived that identified constraints can be removed by providing quality feed and fodder to animals (more nutritious feed and fodder throughout the year) sound health and management practices (good housing, deworming, vaccination) and more suitable animals for cold desert high altitude regions through breeding of the animals to improve their health and productivity.

6. Acknowledgements

All the authors are thankful to Dr. Nazir-Uddin Darokhan, MVSc, Chief Animal Husbandry Officer, Leh for critical suggestions and certain vital information regarding dairy farming in Leh-Ladakh.

References

Dong, S. K., Wen, L., Zhu, L., Lassoie, J. P., Yan, Z. L., Shrestha, K. K., Pariya, D. and Sharma, E. (2009). Indigenous yak and yak-cattle crossbreed management in high altitude areas of northern Nepal: A case study from Rasuwa district. *African Journal of Agricultural Research,* 4: 957-967.

Farooquee, N.A., Majila, B.S. and Kala, C.P. (2004). Indigenous knowledge systems and sustainable management of natural resources in a high altitude society in Kumaun Himalayas. *Indian Journal of Humane Ecology,* 16: 33-42.

ICAR. (2002). Handbook of Animal Husbandry, DIPA, ICAR, Pusa, Delhi-110012.

Sherpa. (1991). Livestock Development in the Himalayas Regions of India, Volume1. Lucknow, India: Society for Himalayas Environmental Rehabilitation and People's Action (SHERPA).

□□□

Chapter-15

Small Ruminant Husbandry Practices in Himalayan Regions

Vijay K. Bharti, A. Biswas, G. Venkatesan, P.B. Deshmukh, P. Kumar, Guru Charan and R.B. Srivastava
Defence Institute of High Altitude Research (DIHAR), DRDO, Leh, Ladakh

1. Introduction

Many local farmers depend on sheep and goat husbandry for their livelihoods. The cold desert area exhibits an extremely cold and dry climate for most of the year which hinders the growth and productivity of biomass. In winter, the temperature falls upto -35°C. The altitude varies from 3048 to 3658 m from mean sea level (MSL). The atmospheric oxygen pressure is 30 percent lower than at MSL that result in hypobaric hypoxia-induced maladies in animals. These type of deflected climatic conditions lead to abnormality in normal breeding which becomes a challenge for animal researchers. The previous technical approaches to livestock production in cold arid Himalayan region of Leh-Ladakh neglected the social, cultural and ecological particularities of livestock production specifically small ruminant production. The research into indigenous sheep and goat management systems is very important to optimize the integration of the traditional livestock management systems developed by local people into modern science and technology and rational policy-based livestock production systems so that the positive impact is maximized and any negative impact minimized (Sherpa, 1991).

2. Prospect of Commercial Sheep and Goat Farming for Meat Production

Whether in cold arid uphill or hot arid desert or hilly tracts of mountains or ravines constituted of leached soil, goats have survived and sustained the poor people. The present worldwide distribution of goats shows that the number of milch type goats are more in the temperate zone and dual type or meat type goats are primarily located in the sub tropical and tropical Asian and African countries. The non vegetarian in India, irrespective of caste, creed and religion, prefer sheep (mutton) and goat meat (chevon) over beef and buffalo meat. Defence Institute of High Altitude Research (DIHAR) has developed broiler sheep and hybrid goat production technology for cold desert of Leh-Ladakh. These animals are well adapted and efficient to this environment and producing good quality tender meat. Local farmers have adopted these technologies for their livelihood by selling of meat in open market and to troops. Therefore, it is necessary to have meat purpose breed of sheep and goat for enhanced meat production in this region.

3. Salient Features of Small Ruminant Husbandry Practices

There are some problems and limitations for optimizing small ruminant production in Leh-Ladakh. Environmental problems and physical constraints such as lack of road connectivity, drinking water in some high altitude summer pastures limit their access to good grazing lands, which results in unevenly utilization of pasture resources. Also, technical weaknesses related to poor animal health care, animal breeding, fodder cultivation and preservation associated with inadequate public services as well as low institutional and human capacity-building limits the sustainable development of sheep and goat production in this region. Many farmers have shifted from livestock production to hotel management or other tourism-related business as tourism has increased many fold in Leh-Ladakh. However, many farmers see the great potentials in small ruminant production for meat. In recent years, local farmers have got significant benefits from central and local governments, DIHAR, Indian troops, NGOs support programs of education development, poverty alleviation and infrastructure improvement, which are not only the driving forces for the improvement of their livelihoods, but also for the development of their livestock production systems.

According to the physical features, climatic conditions, demand for forage and availability of pasture land, the farmers have established different migratory routes and pastures for grazing sheep and goats. The necessity of feed supply for maintenance, movement, growth, production and reproduction required by the grazing animals is one of the essential cause for movements of the herds during the course of the year. The animals are moved towards high hills pastures as well as graze on lower hill pastures in the summer season.

Animal feeds on leaves of willow, poplar tree, seabuckthorn bushes, etc. during summer months. In summer, there are few opportunity of supplementary feeding for the animals, except that they are fed daily with salt to supply the minerals. In lean/winter period, some supplementary feeds, mostly the roughages (dry lucerne) are provided. Many farmers harvest the surplus of forage from summer pastures and agriculture lands and produce hay to feed the goats and sheep in the harsh time (Farooque *et al.*, 2004).

Although the cultivation of forage grasses or plantation of fodder trees for hay conservation is new to most farmers, however some farmers grow lucerne, oat, barley, *etc.* for hay production. The cultivated forages are usually harvested and piled on the roofs of houses for air drying and supplied to the animals as supplementary feeds together with hays or straws in feed deficient winter. The flocks are sometime supplemented with a small amount of concentrates in harsh lean period.

Specialists in veterinary care area are not available in Himalayan ranges. Local farmers are very knowledgeable in disease record, poisonous plant identification and disorder remedies for their flocks. In most cases, local farmers treat the sick animals by traditional healers. Herbal remedies for various diseases, pests and plant poisoning are good examples of indigenous knowledge applied by local farmers. Brisket disease (High altitude disease/Mountain sickness/High altitude pulmonary edema) is frequently observed ailment in poorly adapted animals, where they develop edema due to low oxygen pressure at high altitude that induces laboured respiration, increased heart rate, elevated blood pressure, and fluid accumulation in the thoracic cavity of affected individuals (Sedhain, 1993).

4. DIHAR's R&D Efforts Towards Small Ruminant Production in Cold Desert

Ladakh is a high altitude cold desert characterized by extreme temperatures and harsh climatic conditions which makes agricultural and animal husbandry practices very difficult in this region. As Ladakh is a strategic location from country's defence point of view, large number of armed forces have been deployed in this region. Stressful conditions of this region demands healthy nutrition to the troops for their efficient performance at high altitude. The fresh meat requirement of the troops deployed in Leh garrison is 3146 kg/day. Moreover, civil population of Leh region demands 12000 kg of fresh meat daily. Estimated meat production in this region is only 3000 kg/d. Thus there is 100 percent and 60 percent deficiency of fresh meat availability to troops and civil population in this region, respectively. In view of this, the thrust research areas of DIHAR are identification of suitable meat breeds of sheep and goat and their performance evaluation under different rearing systems in cold desert.

4.1. Broiler Sheep Technology

Due to limited availability of fresh meat locally, procurement of processed and frozen meat from the plain region is cost bearing affair with compromised nutritional and hygienic qualities. To solve this problem DIHAR has evolved meat purpose fast growing broiler sheep by crossing Muzaffarnagri (MUZF) (♀) and Changthangi (CHANG) (♂) sheep. It is a fast growing mutton purpose breed of sheep developed especially for high altitude conditions of Ladakh in order to facilitate round the year availability of fresh meat to armed forces deployed in this sector. It has been developed by crossing Muzaffarnagri female and Changthangi male and scientific assessment of their performance at

Fig. 1 : Adult Broiler Sheep

Fig. 2 : Growing Broiler Lamb

high altitude conditions. This technique leads to improvement in genetical constituents in progeny which provide survivability optima. Their weight gain is 145 g/d up to 8 weeks of age which is superior as compared to other crosses and pure breeds of sheep (Jadhav *et al.*, 2011), (Fig. 1&2; Table 1).

Table 1 : Comparison of offspring's of pure breeds of Muzaffarnagri (MUZF) and Changthangi (CHANG) sheep and their F_1's

Particulars	Muzaffarnagri	Changthangi	F_1 MUZF♀ x CHANG♂	F_1 CHANG♀ x MUZF♂
Average weight (kg) at birth	3.40 ± 0.47	1.85 ± 0.07	3.60 ± 0.60	2.66 ± 0.42
2nd week	4.20 ± 2.54	3.10 ± 0.90	4.15 ± 2.65	3.48 ± 2.40
4th week	5.50 ± 2.11	3.80 ± 1.69	7.06 ± 2.05	4.98 ± 2.04
6th week	6.77 ± 3.90	5.20 ± 3.11	10.13 ± 3.10	5.86 ± 2.62
8th week	9.30 ± 3.68	6.50 ± 4.10	11.73 ± 3.81	7.20 ± 1.96
ADG in bd.wt.(g)	105.36 ± 17.77	83.04 ±25.90	145.18 ±27.74	81.07 ±7.27

Blood biochemical profile indicated that broiler sheep has well adaptation and superior nutrient utilization abilities. Prospective users of broiler sheep are defence forces and military farms based at high altitudes, local farmers, shepherds and farmer's co-operative societies of Ladakh region.

4.2. Local complete feed for sheep

Availability of conventional feeds and fodders for animals in this region is very scarce which is one of the limiting factors in improving animal productivity. DIHAR has developed an economical, balanced complete feed for sheep with the inclusion of local unconventional feed ingredients of this region. It can be formulated as: Wheat straw-20.0 percent, lucerne hay-15.0 percent, tree leaves (admixture of willow and polar tree leaves)-10.0 percent, seabuckthorn leaves-10.0 percent, seabuckthorn pomace-5.0 percent, apricot cake-5.0 percent, soybean meal (solvent extracted)-12.0 percent, wheat bran-4.0 percent, crushed barley grains-18.6 percent, mineral mixture-0.2 percent, and salt-0.2 percent. Feeding experiment with this feed indicated the 10.16 ± 0.74 Feed: weight gain ratio and Feed cost/kg gain in body weight was ₹ 77.06 ± 5.61. These findings clearly indicate the potential of this feed for sheep in cold desert part of world.

Now, DIHAR is working for medicated multinutrients toffee and low cost economic balance ration for goats as neutraceutical/health supplements for better performance of meat purpose goats in this region.

4.3. Development of region specific hybrid goat for meat

Beside this, suitable package of broiler sheep rearing practices have been developed by DIHAR, DIHAR started R&D towards goat production that would further contribute immensely in augmenting fresh meat production locally and throughout the year. Goat meat is very demandable to the army deployed in Ladakh region. The requirement of meat for the local population and the troops has been growing exponentially and the availability of the same was not enough. Among the domesticated ruminants, goats have found their home in the most diverse environment of soil, plant, climate, disease and terrain. The gene-environment interactions over generations and ages have indeed modeled the genetic diversity in Indian goats which are needed to flourish in the plights of quite cold arid, hot arid, hot humid and temperate conditions. DIHAR has introduced many goat breeds for hybrid goat production in cold desert region *viz.* Black Bengal (BB), sirohi (SRH), Jamunapari, Marwari, Barbari, Zakharina, Pasmina and Chegu which were studied for economic and genetic traits.

Though local goat viz. bakharwal (BKW) and CHG breeds were found to have better adaptability and weight gain as well as better reproductive performance in a study conducted during late eighties. The findings from acclimatization/initial adaptation studies revealed that low Lander BB and SRH goats can adapt very well to the harsh conditions of Ladakh. BB is most prolific kidding among Indian breeds. Multiple births are very common, two, three or four kids born at a time, kidding twice in a year with an average age at first kidding is 9-10 months. Meat of this breed is excellent and palatable. Sirohi breed's average live weight of buck is 50 kg and doe is 23 kg and birth weight is 2.0 kg. Kidding is once in a year, twins are common; average age at first kidding is 19 months. Meat of this breed is of very good quality.

These breeds are being crossed (natural service and artificial insemination) with local Changthangi (CHG) goat, which is well adapted to this region (Fig. 3&4). DIHAR has applied classical breeding and selection methods along with molecular aids viz. identification of DNA markers (marker-assisted selection procedures/MAS) for meat and adaptive traits/alleles. These genes and alleles that influence adaptation to specific stresses and challenges will shorten the performance study in progeny and enhance the accuracy with phenotypic studies.

CHG goats are native breed to this region and are well adapted to hypoxic, cold and high altitude conditions of Ladakh. Based on hypothesis that crossbreeding of two different breeds results in enhanced performance, it was decided to perform crossbreeding of BB and SRH with native breeds for further development of region specific hybrid goats for meat and milk purpose. Initial experimental (adaptation and reproductive performance) results are very encouraging.

Therefore, further improvement through crossbreeding and transfer of this technology to suitable entrepreneurs for commercial production of hybrid goats in cold desert region is possible. Hence, supply of good quality tender meat to army through Fresh Supply Depot (FSD) is feasible and this would enhance performance of troops through the availability of nutritious and healthy meat even in severe winter months of the year in Ladakh. Besides this, an increased production of fresh, hygienic meat available locally will improve nutritional status of the civil population of this region and would improve their socio-economic status through employment generation.

Fig. 3 : Hybrid kids (CHG ♂ x ♀ BB) at DIHAR

DIHAR breeding policy for this ambitious project is very novel, which involves finalization of the important traits to be incorporated and developed in the evolved hybrid goat. The selection of the trait depends on the demand of the particular animal product in specific environmental and social conditions. Important traits to be developed and atmospheric constraints in the development of hybrid goat in Ladakh are listed in Table 2.

Fig. 4 : Well acclimatized experimental hybrid kids

Table 2 : Traits of specific importance in goat and sheep and constraints in development of adapted hybrid animals for meat purpose in Ladakh.

Trait to be developed	Adverse effects of the environment on the trait to be developed	Importance of the trait specifically in this region
1.Fast growth	Hypoxic, cold climate hampers growth of the animals.	Required for economical meat production to satisfy ever increasing need for meat to the troops deployed and civil population of this region.
2.Survivability and disease resistance power	Stressful environments conditions increases susceptibility of the animals to pneumonia and further related complications. Also, wear and tear of body tissue decreases the life span of the animals.	Survivability and disease resistance especially in peak winter months will make fresh, healthy and hygienic meat supply available in this region.
3.Weaning weight	Harsh climate adversely affect active growth of the young one.	Increased weaning weight will reduce saleable age of the lambs leading to more profit in short time.
4.Hair/Wool quality and quantity	Prevailing severe sub zero temperature demands warm hair/wool covering for protecting the body which is an important & natural defence of the animal to harsh environment.	It will provide better insulation and protection from severe winter conditions necessary to avoid health complications and subsequent losses.

5.Feed conversion efficiency	Scarcity of the feeds/fodders in this region and fulfillment of specific nutrient requirement as well as ability to digest alpine pastures seek an attention towards the feed conversion efficiency of the animals.	As it is known that nutrition alone accounts for more than 50 percent of the expenditure in livestock rearing, better feed conversion ability in the cold high altitude conditions will further reduce the rearing cost of the animals keeping them healthy.
6.Adaptability to hypoxic conditions	Hypoxic, cold conditions make animal more susceptible to high altitude maladies.	Increased adaptability of the animals to pulmonary oedema, congestive heart failure will reduce financial loss due to subnormal performance of the animals and subsequent mortality.
7.Meat quality	Stressful environmental conditions resulting in changes in body metabolism with respect to protein, energy, fat and minerals affect the meat composition and its quality.	Safe, healthy, hygienic & fresh meat supply to the troops will provide better nutrition and enhance their performance especially in winter.

Underground houses were made for sheep and goat to avoid extreme low temperature and it was experienced that in underground houses the temperature did not go below sub zero.

Table 3 : Goat meat production profile

Breed	Live Weight (09 months) (kg)	Hot carcass (kg)	Dressing (%)	Bone (%)
Sirohi	37.40	21.8	52	13
Black Bengal	10.00	4.47	44	-

The great advantage from the view point of meat production (Table 3) of the high prolificacy of goat is not always matched by fast growth rate. It has however been shown that goats can grow up to 250g/day provided that they are adequately fed. The low growth rate goes with poor food conversion ratio.

5. Farm Management Strategies for Optimum Performance at High Altitude

It is well known that, no animal or breed maximizes the conversion of input to salable product across all environments, nor is the genetic makeup of any animal or breed optimally suited to the challenges encountered in any one environment (ICAR, 2002). The cross breeding can benefit from the higher performance of hybrid vigor (more meat production) and the crosses are better adapted than the parents to various range of altitudes.

The opportunity exists to improve profitability of sheep and goat production for meat through programs designed to achieve balanced genetic potential for adaptation, production and product quality within specific environments of cold desert. Therefore, first we have to characterize important stressors in production environments of this region and identify opportunities to improve adaptation through genetic means. Age at first oestrus is repeatable and heritable. Therefore selection by date of first oestrus will reduce barren goats.

Disease is the major impediment against a successful goatary. Identification and prevention of disease in time are essential features of a healthy security programme. Common diseases in this region are pneumonia, goat pox, brisket disease, cataract, PPR, collibacillosis, brucellosis, abortion and embryonic death of unknown etiology, etc.

Stillbirth and neonatal mortality are the present problems due to cold stress particularly under semi-intensive management conditions. Neonatal mortality can be prevented by improving the level of nutrition in advanced stages of pregnancy (last 6 weeks), ensuring hygienic conditions in the kidding sheds, providing proper bedding and ensuring early feeding of colostrum. Artificial heating (Bukhari/Heat King installation) for two weeks and antioxidant (Vit. E, Selenium, Zinc supplementation) at DIHAR's farm proved to be beneficial in reducing kids' mortality.

6. Conclusion

Small Ruminant production in cold desert has better oportunity to fulfill fresh nutritional requirement. It gives challenges to researchers as climatic conditions challenge the R&D Activities. However, still more research/R&D efforts are needed to develop high productive adapted animals and to produce quality fodder/grazing pasture for economic rearing so that more farmers would be involved in sheep and goat rearing practices.

References

Farooquee, N.A., Majila, B.S. and Kala, C.P. (2004). Indigenous knowledge systems and sustainable management of natural resources in a high altitude society in Kumaun Himalaya. *Indian Journal of Humane Ecology,* 16: 33-42.

ICAR. (2002). Handbook of Animal Husbandry, DIPA, ICAR, Pusa, Delhi-110012.

Jadhav, S.E., Guru Charan, Tilakraj, K., Bharti, V.K. and Singh, S.B. (2011). Performance and blood biochemical profile of lambs fed local unconventional feed ingredients at cold and high altitude conditions of Ladakh. *Indian Journal of Animal Sciences (In press).*

Sedhain, D. (1993). A study of the transhuman system of management of sheep in Rasuwa and Nụwakot districts. In: Tamang, D., Gill, G.J. and Thapa, G.B. (Eds.) Indigenous Management of Natural Resources in Nepal. Proceeding of the Workshop on Indigenous Management of Agriculture and Natural Resources. June 8-9, 1992, Dhulikhel, Nepal. pp 279-289.

Sherpa. (1991). Livestock Development in the Himalayan Regions of India, Volume1. Lucknow, India: Society for Himalayan Environmental Rehabilitation and People's Action (SHERPA).

❑❑❑

Chapter-16

Commercial Poultry Farming in Cold Arid Region of Leh-Ladakh

A. Biswas, Vijay K. Bharti, P.B. Deshmukh,
G. Venkatesan, G.R. Bhagat and R.B. Srivastava
Defence Institute of High Altitude Research (DIHAR), DRDO, Leh, Ladakh

1. Introduction

The North western part of Himalaya shows varied climatic and topographic features. This part of Himalaya exhibits extreme cold and dry climate for most part of the year, which hinders the growth and productivity of biomass. Vegetation cover is very less in areas like Ladakh, Kargil, Drass and Lahaul-Spiti. In winter the temperature falls to -25°C and the altitude varies from 3048 to 3658 m from mean sea level (MSL) in Ladakh. Here, the atmospheric oxygen pressure is 30 percent short of MSL. This type of cold arid and desert like condition does not suit Poultry birds for their natural habitat. However, one non-descript breed of native chicken has been noted in Ladakh whose number is declining very fast.

The term "Poultry" now includes a number of avian species such as chicken, duck, turkey, geese, guinea-fowl, quail, ostrich and emus domesticated for economic purpose. However, chicken accounts for more than 90 percent of the total Poultry population of the country. Chicken and ducks are reared for egg and meat production, turkey, guinea fowl, geese, ostriches and emus are maintained only for meat production. The growth of poultry industry in India has been phenomenal in last two decade, and has reached a stage of self-sufficiency and sophistication. Today Indian poultry Industry ranks 4th in Egg production. However, the picture is very grim in cold arid part of Himalaya. Here still it is a backyard venture with very little number of birds. The non-existence of any commercial hatchery, problem in transportation of Day Old Chicks and unavailability of poultry feed acts as the major obstacle in growth and popularization of poultry farming in this region.

Due to sustained efforts of State Govt., Animal husbandry department, N.G.O'S and few Agro-Animal research organizations, the awareness level and importance of poultry farming among farmers has increased. The presence of Army, increased arrival of tourists, changed life-style and food habits creates huge demand of poultry meat and eggs. Local availability is very less and transportation becomes difficult in winter. On field agricultural activities ceases in winter. So, farmers can go for indoor occupation like poultry farming. In spite of various constraints, it indicates big scope for poultry industry in this region. Few farmers have come up with small scale Broiler farms in cities like Leh and Kargil. At Backyard level most of the farmers opt for 15-20 egg laying or dual purpose birds. Poultry population in Leh-Ladakh district is reported to be 8,267 (Bhagat *et al.*, 2003) to 0.12 lakhs (Phuntsog, 2003). For Kargil district it is 2, 61,171 in the year 1999-2000 (Gupta, 2000).

2. Major Constraints and Difficulties in Poultry Farming in Cold-Arid Region

To become successful in popularization of poultry farming in cold-arid region, it is imperative to understand all constraints and difficulties which comes in way of poultry production which are described below:

2.1. Religious Sentiments

Major population of this region is Buddhist community. They are vegetarians and avoid killing of animals. Their interest in poultry is limited to keeping egg laying birds.

2.2. Extreme Cold and Hypoxic Environment

A day old chick (DOC) of any species of poultry is physiologically prone to adverse effects of cold stress because of their poor homoeothermic limit. The temperature range for optimal production is 17-23°C (Thermo-neutral zone) only. Any variation of temperature above or below this range affects the growth and production parameters of birds. In this region the temperature fluctuates from + 35°C in summers (day temperature) to -25°C in winters (night temperature) which exerts huge stress on homoeothermic control system of birds as well as huge expenditure on maintaining the optimum temperature inside the house in extreme winter. The altitude varies from 3048 to 3658 m which directly affects the hatchability percentage of chicks because of reduced partial pressure of oxygen in air. This single factor is mainly responsible for non- existence of any hatchery in this region and in turn non-popularization of poultry farming. Cold stress coupled with hypoxic and dry environment (Relative Humidity (RH) =35-45%) reduces the optimum production and livability potential of poultry birds.

2.3. Unavailability of Day Old Chicks and Poultry Feed

Due to non-existence of hatchery and poultry feed shops in near by places, it becomes difficult to procure DOC and ready made poultry feed in winter from other parts of the country; when road route becomes blocked and the region remains cut off for at least 6 month. Difficulty of transportation in high altitude mountains and high transportation charges adds to the burden of farmers. Local availability of poultry feed ingredients is minimal and there is hardly any source to get them at regular interval at local level. Unavailability of suitable poultry breeds adapted to this region with optimum production potential is another problem.

2.4. Suitable Germplasm

Considering the climatic condition and constraints, selection of appropriate germplasm for broiler and layer becomes very important. Germplasm should be able to withstand harsh climatic condition and produce optimally. Research findings of DIHAR, Leh, DARL (DRDO), Pithoragarh and J&K state Animal Husbandary (AH) department suggest Belicharana Brown breed and commercial egg laying strains of White Leghorn (WLH) and Road Island Red (RIR) are good for egg purpose. These two breed survive well with good production performance even in backyard of farmers. F_1 strains of slow growing commercial broilers are suitable for backyard level of farming. In poultry houses with better temperature control, fast growing Venky broilers have been found more economical by virtue of their rapid body weight gain and lesser Reed Conversion Ratio (FCR). But in general, cockerels of slow growing strains (Chabro, Central Poultry Breeding Farm (CPBF), Chandigarh) and dual purpose Indian varieties like Dhanraja, Vanraja will be more appropriate because of their hardy nature and better survivability at farmer's level. Average body weight of Vanraja chicks at day old and at 6th week of age is 37.97 ± 0.60 g and 810.13 ± 29.60 respectively at the altitude of 5200 ft. above MSL. The average FCR was 2.64 ± 0.04 kg at 6th week with 22.63 percentage total mortality during this period. This shows that the growth of broilers at high altitude is lesser compared to plains and they consume more feed per unit body weight gain. To encourage poultry farming at backyard level there is need to introduce dual purpose broody Desi birds. Till the time hatchery is established or facility for Day Old Chicks (DOC) is made available, at least farmers can perpetuate backyard poultry keeping by getting chicks hatched from Desi brooding birds even in winter.

Other species of poultry which can be reared in this region (as reported by research findings of DIHAR and Department of Animal Husbandry, Leh) are Khaki Campbell (layer), Super Vigova (broiler) a Duck breed, Japanese quail, Emu, Ostrich, Guinea fowls and Turkey. Wild varieties of duck (*Anas platyrhyanchos*), quail (*Coturnix coturnix*) and goose (*Anser anser)* have been reported to thrive in wild areas of this region (Ladakh). These hardy birds show good tolerance to harsh climate.

In-spite of the constraints of hatching and feed unit, there is a tremendous demand of poultry in Leh-Ladakh.

3. Poultry Housing System in Cold Arid Climate

The housing of poultry birds is difficult in cold arid climate especially in winter. To maintain the optimum temperature of 25-35°C inside brooder and 15-20°C inside grower/layer house, special care, accessories and some modifications in design of houses are required. The systems of housing prevalent in this region are semi-intensive to backyard. The types of houses used by farmers are given below.

3.1. Over Ground Mud Walled Poly Shed

These sheds are made of double wall mud bricks, with the inner first third up from the floor consisting of cement. The roof is made up of bamboo, wood, grass and other locally available materials. The middle third of one side of the wall facing the sun is made from a polythene sheet supported by wooden logs and at the centre is a glass window. This whole portion is then covered with another white polythene sheet to trap the sun light and thus maximize the heat inside the house. The inside half of the wall is covered by blankets and plywood to preserve heat. False roofing has been attempted using a thick woolen sheet at half the height of the house. At the centre of the house, there is local Bukhari (wood burning stove), fitted with a chimney and wrapped with wire mesh to prevent the birds coming too close. In the peak of winter in the L-sector of Leh-Ladakh, when the outside temperature falls to -4°C during the day, the temperature inside at the bird's level has been noted to be +8°C without lighting the Bukhari (Fig. 1-4).

Fig. 1 : Over ground mud wall poly shed

Fig. 2 : Blanket and ply wood covering walls

Fig. 3 : Inside view

Fig. 4 : Brooding of chicks by farmer

3.2. Semi-Underground Mud /Stone Wall House

These houses are used to protect the birds from the extreme cold and comprise different types of underground and semi-underground sheds, made up of mud, mud brick, stone and concrete. Some of these houses include extra poly-houses as an extension to allow the bird's access to direct sunlight during the day-time. These semi underground sheds are constructed of a double wall of mud or stone up to the ground level, and mud bricks being used to construct the rest of the wall as well as the roof. The roof is double layered with mud bricks and part of ceiling inside is supported with wooden logs and locally available packing materials. False roofing is made by covering the inner aspect of the roof with polythene, cardboard and paper or ply wood (Fig. 5-8).

Fig. 5 : Mud brick roof with stone wall

Fig. 6 : Wire, glass & polythene made Extension

Fig. 7 : Farmer inside the semi-underground poultry shed

Fig. 8 : Inside view of semi-underground poultry shed

One or two windows are made where the wall interfaces with the ground. These are either fitted with wire mesh or adjustable wooden or glass panels opening outside, or covered with curtains or polythene. The floor, corners and lower one third of the walls are cemented for easy cleaning and protection from rodents.

It has been reported that mud bricks are better insulators than stone. These houses are cheap to build and are affordable to farmers. Another type of semi-underground house has been attempted by the Defence Institute of High Altitude Research (DIHAR). In this instance, the walls and floor were made of concrete bricks and asbestos sheets were used for the roof. At the centre of the roof, polycarbonate sheets were used to trap solar radiations. A double door was used to prevent incoming air drafts. In both types of houses, one opening is left in the roof for the chimney of the Bukhari.

3.3. Trombay Wall Shed

This housing is a scientific adaptation of typical local poultry houses; designed to cater to the needs of farmers by the DIHAR. They are made of a double wall of mud bricks which are plastered with mud on all sides (Fig 9).

Fig. 9 : Trombay wall shed

This above-ground shed has two windows and both the walls and floor are cemented. The lower one third of the walls is covered with galvanized aluminum sheets for easy cleaning. The roof is the same as the farmer's other sheds. An extra door is located at right angles of the first door. Some other modifications have also been attempted, such as including large double layered glass windows on the whole face of the wall facing the sun. The outside of the walls is painted black and a polycarbonate sheet is located at the centre of roof. Many small windows are fitted on the lower part of the walls in order to meet ventilation requirements. As per requirements, window doors open partially or fully to allow ventilation. All of these modifications augment the solar heat trapping capacity of the housing and improve the ventilation of the house. This modified housing can be easily constructed with locally available materials and is cheaper to build. The results of experimental trials conducted by the DIHAR for growth, (FCR) and mortality of broilers in this type of house were satisfactory, even in peak winter conditions.

3.4. Solar Poultry House

The solar house has been designed by DIHAR to provide maximum comfort, maximum protection from the cold, better ventilation and hygienic conditions for the birds (Fig 10, 11). This type of housing is very costly but most efficient in maintaining optimum temperature and ventilation inside (Fig. 10&11).

Fig. 10 : Outside of Solar Poultry House

Fig. 11 : Inside of solar poultry house

4. Feeds and Feeding

Poultry birds are monogastric and their feeds consist of mostly grains. Food grain production is limited in this region. Farmers are provided subsidized human ration through Public Distribution System (PDS). But for animal ration there is no such system and food grains/ poultry feed have to be imported from some where else. Poultry farmers face three major hurdles in feeding their birds.

4.1. Limited Availability/Unavailability of Poultry Feed/Food Grains

Wheat, Barley, Pea and Mustard are the few major local food grains, which are produced by farmers in limited quantity. Very little amount of surplus grains are available with the farmer to spare for poultry. They import either poultry feed ingredients like maize, soybean and wheat bran or purchase ready-made feed from other parts of country when road route remains open i.e. summer only. Transportation and storage of feed from distant place exerts extra economic pressure.

4.2. Unscientific Feed Formulation/Improper Feed Formulation

Balanced readymade feed is not always procured by farmers. To save money, they purchase few ingredients and supplement if with locally available feeds. This unbalanced feed decreases the weight gain in birds and increases feed conversion ratio. Ready made starter feed is given to chicks and maize based locally prepared poor quality feed is provided to birds above 4th week of age. In winter at the time of scarcity of feed, farmers make their own poultry feed with locally available vegetables, cereals, and some other things. This includes spinach, cabbage, onion, garlic, lucerne leaves, crushed turnip, barley, gram, wheat flour and mustard oil preparation.

4.3. Need to Standardize Energy and Protein Percentage Level in Feed

Scientific feed formulation with optimum Metabolizable Energy (ME) (K.cal) and Crude Protein (CP) % is very essential for this region. Because of cold stress and high altitude, the ready-made feed available in market would not be appropriate for birds of this region. Ascites becomes a major problem in fast growing broilers if energy and protein level in feed is not optimized. Depending on season (winter/summer), quality and type of day old chicks (fast/low growing), scale of feeding (Ad-lib/restricted) and incidence of ascites, feed formulation needs to be adjusted and energy and protein level along with feed additives has to be optimized.

5. Scientific Feeding of Poultry for Optimum Production in Cold Desert

Feeding of poultry constitute major managemental practice in poultry production since feed constitutes 60-70 percent of total expenditure. In cold arid hypoxic environment of Himalaya, poultry birds attain lesser body weight gain and consume relatively more feed to meet energy needs for endogenous heat production. On the other hand excess feeding leads to Ascites and poor Feed Conversion Ratio (FCR). Therefore feeding scale and nutrient density of the feed for Broiler and Layer should be selected appropriately. The major nutrients required for poultry are: (a) Water (b) Proteins (c) Carbohydrates (d) Fats (e) Minerals and (f) Vitamins (g) Feed additives.

Carbohydrates and fats are principle source of energy. Fats provide 2.5 times more energy than carbohydrates. But the digestibility of fat and percentage inclusion in poultry ration is very low (4-8% only). Protein is essentially required for body growth and for supply of essential amino acids which are not synthesized in required amount in body. Poultry requires arginine, glycine, histidine, leucine, lysine, isoleucine, methionine, phenyl alanine, threonine, tryptophan and valine as essential amino acid, out of these arginine, lysine, methionine, cystine and tryptophan are critical amino acids, since they are usually deficient in poultry diet. Lysine and methionine are considered as limiting amino acid and must be included in diet. Cereal grains are usually low in critical and limiting amino acid. Thus it is must to include a good proportion of animal proteins to ensure the inclusion of all critical amino acids in diet. Minerals and vitamins do not supply energy but regulate several essential metabolic processes in the body and in their deficiency several disease conditions arise in poultry. Essential minerals and vitamins required for poultry are calcium, phosphorus, sodium, copper, iodine, iron, maganese, zinc, selenium, vitamin A, vitamin D_3, vitamin E, vitamin C, pyridoxine, riboflavin, panthothenic acid, niacin, folic acid, vitamin B_{12} and choline respectively. Balanced poultry ration should include all the above nutrients in right proportions for maintenance and productive functions for different age group of birds.

6. Managemental Practices for Broiler Production in Leh-Ladakh

Broiler farming is an intensive operation which starts from procurement of quality chicks to marketing of uniform sized flock. All in and all out system is most commonly practiced in which chicks of same age, weight and of same hatchery are housed together and are reared till the age of marketing and all birds are sold at the same time. Some basic points in broiler production are as follows.

6.1. Procurement of Day Old Chicks

All procured chicks should be disease free or be vaccinated. At least parent stock should be free from Mycoplasma. Only quality chicks of uniform weight, healthy in appearance and mobility should be purchased. The success of Broiler production depends a lot on quality and disease free status of DOC. DIHAR advocates use of slow growing breeds as it is beneficial in this area especially in winters, as has been established by research findings. All procured chicks should be from same hatchery.

6.2. Housing

At a time, chicks of same age group should be kept on farm or chicks of same age group/same breed should be housed in a shed. If birds of two different age group / breed are to be housed, there should be sufficient distance between two sheds. Houses should face south-north to catch Sun rays especially in winter. They should be at leeward side of mountains and sheltered from direct wind. The type of house and housing materials required for maintaining optimum temperature and ventilation depends on the farmer's affordability. The room should be properly ventilated in day time, especially in winter to get rid of ammonia. The poultry house should not have large rooms and ceilings should be low. This facilitates easy warming and ventilation. All cemented parts and windows should be cleaned and washed with any disinfectant (phenyl, attack, aldipol). After drying of house, the house should be fumigated with formaldehyde and potassium permanganate in 2:1 ratio (40 ml: 20 g/100 cubic feet) 24-48 hrs before arrival of chicks and all windows and openings should be sealed.

All windows should be opened 3-4 hrs before placing chicks to remove all the toxic fumes and gases. Waterer and feeders should be cleaned in advance and should be moisture free. The litter material should be fresh and well placed in side brooder mixed with quick lime. Newspaper, waterer, feeder and *Bukhari* should be placed and arranged properly at least 3-4 hrs before arrival of chicks to bring the room temperature at optimum.

6.3. Brooding of Chicks

Brooding is providing supplementary heat to day old chicks until they no longer require exogenous heat (0-6 weeks).There are two general systems of brooding : (1) Natural method, and (2) Artificial method. Natural brooding is done by Desi hens/mother of chicks. Before arrival of chicks in brooder house, the house should be prepared in advance. The broody hen will provide all the warmth needed by the chicks. In a 2 feet square and 2 feet high packing box or wooden basket a deshi hen can brood 15-20 chicken. Small amount of chick feed mixed with water is kept under the box. Artificial brooding is handling of chicks without the aid of hen.

Brooding period for Broilers is 2-3 weeks and for Layers it is 6 to 8 weeks. Brooding period for broilers may go up to 6 weeks depending on weather condition or at least during night. This period is very critical in life of Birds because maximum mortality occurs in this period. Ventilation of brooder house should be judiciously done. There should be a balance between required room temperature and adequate ventilation. At least some windows should be kept open always. There must not be too much moisture and dusty environment inside house. The litter can be of saw dust/paddy husk/coarse wheat bhusa etc. Its thickness may be 5-10 cm depending on whether condition. Litter should be stirred and cleaned frequently to prevent caking. Wet litters if any should be replaced immediately with dry litter. When moisture is low, dust becomes a problem and when high the litter becomes cacked which gives ammoniacal odour and is a potent source of coccidiosis. A good litter adheres slightly and breaks up when dropped from hand.

6.4. Brooding Temperature

Chicks are very susceptible to low temperature (cold stress) and fluctuation of temperature in early stage of their life. Too high or too low temperature slows down growth and causes mortality. During first week the temperature should be 95°F (35°C) which may be reduced by 5°F per week during each successive week till the temperature reaches 70°F. The temperature inside room should be 20°F below brooder temperature. A maximum and minimum thermometer should be placed in brooder house to monitor the room temperature. Behavior of chicks like huddling in corners or coming very close to heat source tells about the temperature inside room /brooder. If chicks are evenly spread under hover/room, the temperature shall be considered optimum. When temperature is less than required, the chicks try to go closer to the source of heat and when temperature is too high, they scatter away from heat. If any draft air is coming then also the birds stray away heat and may be found sitting in corner or any one side of house. To provide artificial heat, hover type electric brooder is most commonly used. But if there is problem of electricity the gas brooder or kerosene based *Bukharies* can be used. In this region both electricity and availability of gas is the problem, so the most common source of heat is Bukhari. Regular check up, cleaning and refueling of Bukhari is must to avoid accidents due to fire and leakage of Kerosene oil.

The brooder space per chick is 7sq inch and if brooding is being done in whole house then the floor space required is 0.1 sq m till the age of brooding. If hover type brooder is used then the chick guard should be placed 60-70cm from the edge of hover. In cold arid region, major mortality occurs in first 1-2 weeks of age because of very low temperature and less humidity. To balance the required temperature and optimum ventilation with *Bukhari* becomes difficult. So, judicious opening of windows at frequent intervals and nights is required. Poultry birds require the same standard of clean water as human beings. The water must be cleaned with potassium permanganate at least once in a day. All left over water of previous day should be discarded and only fresh water be given each time. Luke warm water may be given in winter. Height of waterers should be maintained at 2.5cm above the back height of the chicks to reduce spilage. For the first 2 weeks, 50 cm of linear water space for 100 chicks shall be provided.

The requirement of feeder space can vary from 2.5 to 3 cm per bird from 0 to 8 weeks of age. Chicks should be fed only after 3 to 4 hours after arrival in the brooder. To prevent wastage of feed the feeder should not be filled more than one -third at any time. Like waterer, the feeder equipments should be cleaned once daily. The waterer and feeder should be arranged in a circle around the hover or *bukhari* and they must be placed at equal distance, so that all chicks get equal opportunity for warmth, feed and water. As a thumb rule, the number of feeder and waterer should be such that all chicks must find space to take the water and feed at a time. With advancement of chicks age, number of waterer and feeder must be increased.

6.5. Lighting

Artificial lighting is very essential for broilers throughout the night for first 02 days to allow them to locate feeder, waterer and get organized. Uniform distribution of light in shed makes the chicks to get scattered uniformly and receive little warmth from light source. Electric infrared bulbs or lantern may be used for lighting. Electric bulbs in large number, if hanged one foot above the chick's height gives warmth and chicks prefer to huddle under the light source.

After 2 weeks of age, regular monitoring of broiler's health is required. The condition of litter and amount of ventilation becomes very important. At this age relaxation in temperature control may be permitted because of developed thermoregulatory mechanism of birds. But birds require more ventilation and better hygiene condition. The diseases like Air sacculitis, Chronic Respiratory Diseases (CRD), mixed infections of Mycoplasma and *E.coli* starts manifesting their symptoms at this period. Also a chance of entrance of Coccidiosis is very likely at this period. So the upper layer of litter should be removed/cleaned twice a week. Birds drooping in litter should be checked for coccidiosis. After 3rd week, birds start developing Ascites like syndrome. Increased ventilation is the simplest answer to this problem. To prevent entry of infection, only one person should be allowed to enter inside the house and even that with changed slipper/shoes. One shoe/slipper especially for that shed should be kept separately. In this region no vaccination is required for broilers. Only the hatchery born disease is the source of infection. If severe case of out break is noticed then only vaccine is required.

7. Grower Management

Most of the managemental practices *viz.*, housing, feeding, watering, sanitation and disease control are same as in broilers. The difference lies in intensity of care and supervision. As the birds have attained 08 weeks of age, there is less need to maintain the room temperature. Floor, feeder and water space should be increased. Floor, water and feeder space recommended for a grower are 800-1000 sqcm, 1.5-2.5 cm and 6.5-10.0 linear centimeters respectively. Care should be taken for cleaning of litter and not allowing accumulation of ammonia gas inside house. Duration of lighting should be reduced in decreasing order 08 hrs of artificial light is needed for growers. Increased light intensity leads to cannibalism and vent picking. Vaccination may be required for chicks which are meant to be kept as grower and layer for specific diseases prevalent in that locality.

8. Layer Management

The flock should be transferred to layer house at 18-20 weeks of age. In cold climate it is preferred to keep the flock on deep litter. One laying nest for every 4 pullets is required. The floor space recommended for layer is 1800-2000 sqcm, feeder space of 14 cm and water space of 2.5 cm lighting influences eggy production. Therefore, lighting of layer house is very essential for optimal production. From 21 weeks onward the lighting should be increased gradually till it reaches 16-17 hrs per day and at that level thereafter. Irregular light results in drop in egg production. One 40 watt bulb with a reflector hanged 2.1m above the floor for each 9.29 m square of floor space would provide the recommended intensity of light. Light bulbs should be cleaned regularly. Laid eggs should be collected immediately from the nest in the morning to prevent the layer from eating its egg. Male breeder birds can be kept together or separately. They should be given extra calcium, manganese and vitamin E to ensure proper fertility.

9. Hatchability at High Altitude Cold Region

Despite the current interest in the hatchability performance of poultry eggs, there is scarcity of literature describing physiological processes and adverse changes involved with acclimatization to high altitude in eggs. There are few studies that described adverse effects and physiological adaptation in eggs to high altitude and cold environment. The environment at high altitude is characterized by many physical changes, but the most evident of these changes is the decrease in the partial pressure of oxygen. Hatchability percent per se is adversely affected by hypoxia of high altitude (Lundy, 1969). The environment at high altitude is characterized by many physical changes, but the most evident of these changes is the decrease in the partial pressure of oxygen. Research has shown that hatchability of chicken eggs is reduced as the altitude at which they are incubated is increased. The cause of higher embryonic mortality has been reported to be lower hemoglobin level.

Chicken requires adaptation period for normal fertility and hatchability at high altitude. Hence, if chicks are grown from day old to on set of lay at high altitude, then the survivors will show better hatchability compared to adult hens brought from sea level.

At high altitude the oxygen availability is low and therefore oxygen supply to the eggs embryo is adversely affected. Due to this reason 75-85 percent eggs do not hatch at high altitude in spite of better hatchery managements. Also, poor hatching rate may be due to increased diffusion coefficient of gases at reduced air pressure and this may lead to desiccation of eggs and low survivability of embryos. Therefore, we started study on hatchability at high altitude to enhance local production of day old chicks.

In this direction, DIHAR established a hatchery unit at HQ, Leh in 2010 and conducted pilot studies with CPDO (Central Poultry Development Organization (Chandigarh) eggs. All the eggs were kept for two days in a dark room for acclimatization. Then eggs were set at incubator where temperature (98.8-99°F) and humidity (88-90%) were maintained with the help of auto turning/hour. The pilot study results are as follows:

Table 1 : Pilot study of egg hatchability at high altitude

	First batch	Second Batch
Eggs Set	118.00	90.00
Fertile Eggs (%)	64.00	83.00
Hatching (%)	19.00	29.00
No. of Day old Chicks (DOC)	22.00	26.00
Body Weight of DOC (g)	29.22	30.00
Mortality (%) after 45-Days	18 .00	19.00
Body Weight after 28 days (g)	424.00	420.00

Sources: DIHAR data.

10. Important Prophylactic Measures of Broiler Birds at High Altitude (Cold Arid Region)

Day 1 to 3	: Electrolyte powder, Tylosin tartrate and anti-stress/adaptogens in water. Check for blocked cloaca and give 5 ml of liquid paraffin per 10 liters of drinking water if required.
Day 4 to 6	: B complex, Vitamin A, D, E & K and Liver tonics.
Day 7	: Lasota vaccine as eye drop.
Day 8 to 9	: B complex, Vitamin A, D, E & K. Check for blocked cloaca and give 5 ml liquid paraffin/10 liters of drinking water if required.
Day 10 to 15	: Enrofloxacin 10%orally + B complex + Liver tonics.
Day 16 to 20	: B complex + Vitamin A, D, E & K + Liver tonics.
Day 21 to 23	: Coccidiostat + Vitamin A, D, E & K.
Day 24 to 27	: B complex+ Vitamin A, D, E & K + Liver tonics.
Day 28	: Lasota vaccine in drinking water.
Day 29 to 35	: B complex + Liver tonics.
Day 36 to 38	: B complex + Liver tonics, Pefloxacin orally.
Day 39 to 45	: B complex + Vitamin A, D, E & K + Liver tonics.
Day 46 to 48	: Coccidiostat + Vitamin A, D, E & K.
Day 49 to maturity	: B complex + Vitamin A, D, E & K + Liver tonics.

References

Bhagat,U.C.(2003). Proceedings of workshops on "*Devising strategies to make Animal- Husbandry & Poultry farming sustainable and economically viable in Ladakh*". FRL , Leh , 27 sept 2003.

Gupta, A.L. (2000). Proceedings of workshops on "Devising strategies to make Animal- Husbandry & Poultry farming sustainable and economically viable in Ladakh". FRL , Leh , 27 sept 2003.

Lundy, H. (1969). In: *The Fertility and Hatchability of Hen's Egg*, 143-176. Edit. Carter, T.C. and Freeman, B.M. Edinburgh, Oliver and Boyd.

Phuntsog, T. (2003). Proceedings of workshops on "Devising strategies to make Animal- Husbandry & Poultry farming sustainable and economically viable in Ladakh". FRL , Leh , 27 sept 2003.

□□□

Chapter-17

Zanskar Ponies :
Packers and Movers on the Rugged Terrain

G. Venkatesan, P.B. Deshmukh, A. Biswas, Vijay K. Bharti and R.B. Srivastava
Defence Institute of High Altitude Research (DIHAR), DRDO, Leh, Ladakh

1. Introduction

The Zanskar Range is a mountain range in the Indian state of Jammu and Kashmir that separates Zanskar from Ladakh. Geologically, the Zanskar Range is part of the Tethys Himalayas, an approximately 100-km-wide synclinorium formed by strongly folded and imbricated, weakly metamorphosed sedimentary series. The average height of Zanskar Range is about 6,000 m. Communication with the neighbouring Himalayan areas is maintained across mountain passes or along the Zanskar River when frozen. These topographical features explain why access to Zanskar is difficult from all sides. with most of the remote locations being less connected by roads, the beasts of burden still continue to be the major means of transport in these regions. Relevance of the horse to mankind is well documented throughout the history of civilization. The horse has served the mankind as a source of food, power for agricultural, commercial and military pursuits; past time recreation and sports. Equines have been playing significant role in defence forces since time immemorial. Even in today's era of modernization the equine draught power cannot be wished away for many strategic reasons. This point has been amply demonstrated during Kargil operations (Op Vijay).

2. Home Tract of Zanskar Ponies

Zanskar is a high altitude semi-desert lying on the Northern flank of the Great Himalayas Range. This mountain range acts as a climatic barrier protecting Ladakh and Zanskar from most of the monsoon winds, resulting in a pleasantly warm and dry climate in the summer. Rain and snowfall during this period are scarce, although recent decades have shown a trend towards increasing precipitation. Most of the precipitation occurs as snowfall during the harsh and extremely long winter period. Parts of Zanskar valley are considered some of the coldest continually inhabited places in the world (Fig. 1).

Zanskar covers an area of some 7,000 square km at a height of between 3,500 and 7,000 meters. It consists of the country lying along the two main branches of the Zanskar River. The first, i.e. Doda, has its source near the Pensi-La (4,400 m) mountain-pass. The second branch is formed by two main tributaries known as Kargyag River, with its source near the Shingo La (5,091 m), and Tsarap River, with its source near the Baralacha-La. To the north-east lies the Zanskar Range, this separates Zanskar from Ladakh.

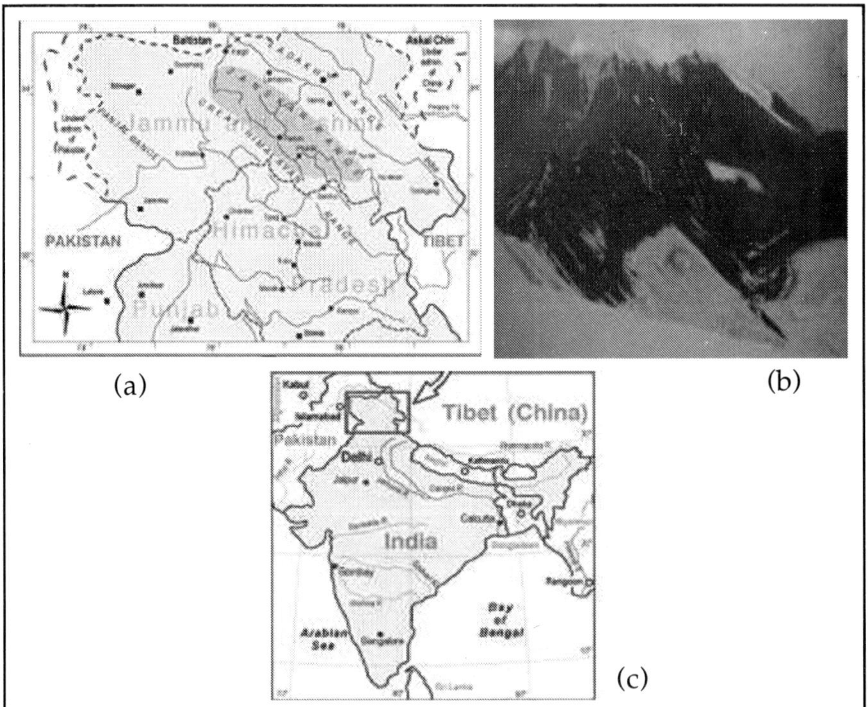

(a) (b) (c)

Fig. 1 (a-c) : Zanskar Range in Leh-Ladakh

These topographical features explain why access to Zanskar is difficult from all sides. The easiest approach leads from Kargil through the Suru valley and over the Pensi La.

Fig. 2 : Home track of Zanskar pony in winter

Fig. 3 : Home track of Zanskar pony in summer

Fig. 4 : Home track of Zanskar pony

3. Breed Characteristics of Zanskar Pony

Zanskar ponies are one of the important indigenous horse breeds notified in the government of India gazette which are found in Leh-Ladakh areas of Jammu & Kashmir (Fig. 2,3&4). DIHAR studied phenotypic and other body characteristics of these ponies and characterized the breed. The common body colours are grey, black and copper. The height ranges between 132–147 cm. The tail is long and heavy, almost touching the ground and body hairs are fine, glossy and long. They are a sturdy breed and have very good adaptability to extreme climatic conditions of high altitude areas of Ladakh.

These animals are generally docile, easy to handle and less prone to diseases of high altitude cold regions. The animals are alert, sure footed, well build, medium in size, have wide face, uniform gait and predominant eyes and load carrying capacity is almost same as that of a mule GS used by the army in this region. But the feed consumption is less as compared to the mule GS. Also these animals can survive on locally available feed and fodder. In case of an urgent/emergency deployment these animals need not undergo the acclimatization procedure. The local population and troops of Ladakh scouts deployed in this region prefer these ponies as compared to mule GS.

4. Military and Civil Utilities of Zanskar Ponies

Zanskar ponies, one of the five recognized indigenous equine breeds of India, had proved their usefulness as pack animals in mountainous high altitude regions of Ladakh. These horses are used for draught, transport, riding and polo. These horses are known for hardiness and ability to work at high altitude. Local people use them as drought/carrier animals for pulling of cart, riding, load carrying, etc. Considering their fast depleting number coupled with dilution of the genetic purity of this breed due to indiscriminate breeding with other nondescript equines and also because of its defence significance. A project was undertaken for exploitation of this local equine germplasm to meet army's requirement of ponies and mules since Jan 2002.

Fig. 5 : Riding and load carrying on Zanskar ponies

4. DIHAR Efforts towards Breed Improvement and Health Managements of Zanskar Pony

The Zanskar Research Station (Pony Section) under Animal Science Division, DIHAR (Defence Institute of High Altitude Research) had already made significant contributions by line breeding of selected parent stocks, studying the reproductive performance of Zanskari ponies and the growth pattern of Zanskari foals.

The Zanskar Pony Section in consonance with the objectives of improving the equine draught power has inducted a Halflinger stallion and a Donkey stallion from Remount Veterinary Core (RVC's). Equine Breeding Studs, Hisar and Babugarh, U.P., respectively for cross breeding and mule production. Establishment of an equine Artificial Insemination Laboratory is another ambitious and futuristic activity this division has undertaken. Zanskar Pony Section also is giving training to the progeny from the ongoing breeding operation for subsequent issue to Ladakh Scout and Indo Tibetan Border Police (ITBP). Pony Training Track of 2.5 km length for training the animals has also been made on a hillock adjoining the Defence Institute of High Altitude Research (DIHAR).

Fig. 6 : Zanskar Pony Research Center (Below) and Training Track at DIHAR

Presently, crossbreeding of these ponies with Italian donkey stallion and Austrian Haflinger stallion is being carried out and the performance evaluation studies of the crossbred offspring's are being studied in DIHAR (Table 1).

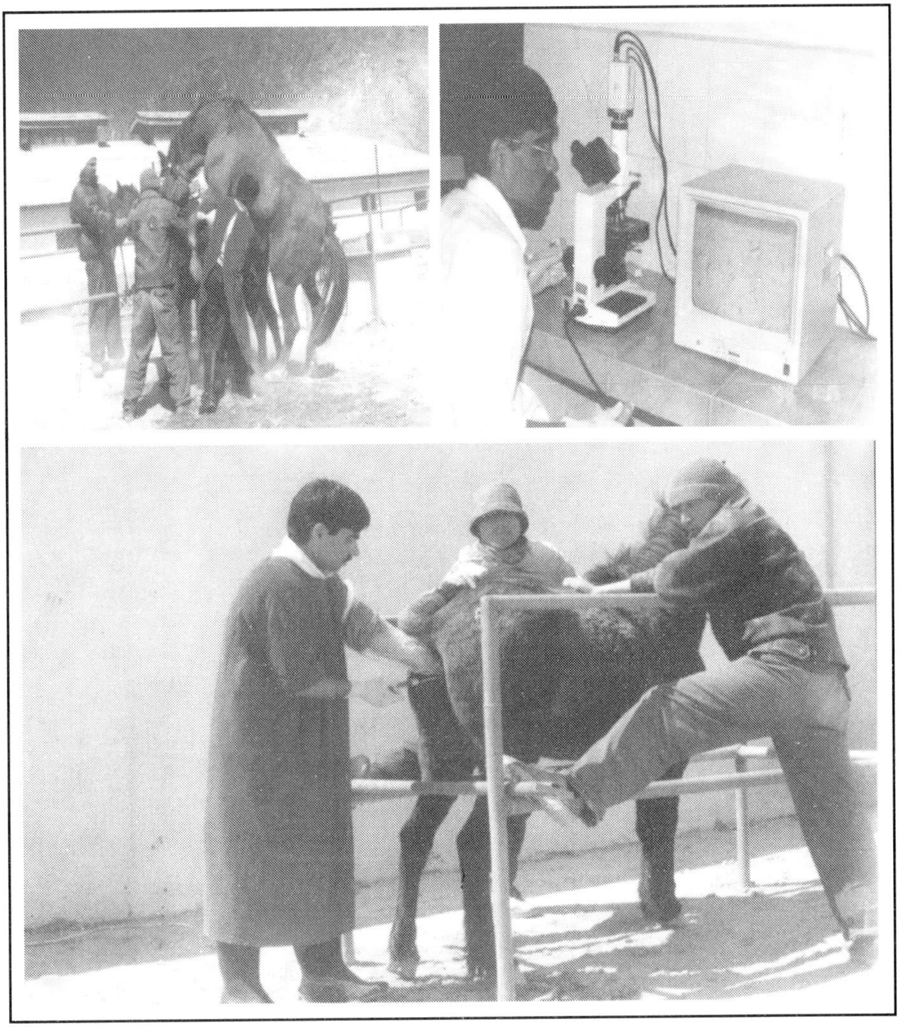

Fig. 7 : Semen collection, evaluation and Artificial Insemination in Zanskar ponies

Zanskar breeding base has been established in Nubra Valley for breeding and training of Zanskar ponies and further issue to Army to meet their requirements. *In-situ* upgradation of Zanskar ponies breed was achieved. Hence, these trained ponies can be deployed for logistic requirements of operational conditions in Ladakh.

Table 1 : Comparative study of blood hematology in adult cross-bred (ZN x HL) and Zanskar male ponies (age 3-Years)

Parameters	Pure Zanskar	Cross-bred
WBC ($10^3/\mu l$)	6.71	6.11
Lymphocytes (%)	58.01	56.81
Monocytes (%)	6.01	5.12
Granulocytes (%)	35.98	38.07
RBC ($10^6/\mu l$)	7.01	8.48
Hb (g/dl)	10.45	12.63
HCT (%)	28.98	35.36
MCV (fL)	41.5	43.68
RDW (%)	18.27	18.98
MCH (pg)	14.9	15.25
MCHC (g/dL)	36.07	35.01
PLT ($10^3/\mu l$)	64.00	73.33

Source: DIHAR, Annual Report (2010).

5. Important Considerations for Feeding Management of Ponies

The diet selected for different horses will depend on the nutrient requirements of particular classes of horses and the combinations of ingredients chosen to meet those requirements. Varying combinations of ingredients may be chosen depending on nutrient content, availability, price of ingredients and preference of the horse owner. All diets for horses should contain adequate amounts of roughage. Thus, all diets must contain either pasture or some type of harvested roughage hay or other forms of roughage (NRC, 1980, 1989).

Dehydration through sweating results in the loss of both water and electrolytes. Therefore, feeding and watering of ponies should be as per their optimum requirements (Table 2). Varying the proportions of roughage and concentrates can be used in the management of horses to control energy intake, to maintain normal digestive tract fill, to minimize digestive dysfunction and to regulate consumption of feeds by horses that are fed in groups. However, it is highly recommended that all horses either have access to a pasture or be fed sufficient long roughage in the diet to minimize the digestive dysfunction often attributed to feeding large amounts of concentrates. A good rule of thumb is that horses should be fed at least 1 percent of their body weight/day of good-quality roughage or be given access to a pasture for sufficient time to allow them to consume at least 1 percent of body weight of dry matter/day.

Table 2 : Daily feed requirements for different classes of horses (% body weight)[a]

	Forage	Concentrate	Total
Mature horses			
Maintenance	1.5-2.0	0-0.5	1.5-2.0
Mares, late gestation	1.0-1.5	0.5-1.0	1.5-2.0
Mares, early gestation	1.0-2.0	1.0-2.0	2.0-3.0
Mares, early lactation	1.0-2.0	0.5-1.5	2.0-2.5
Working horses			
Light work	1.0-2.0	0.5-1.0	1.5-2.5
Moderate work	1.0-2.0	0.75-1.5	1.75-2.5
Intense work	0.75-1.5	1.0-2.0	2.0-3.0
Young horses			
Nursing foal, 3 months	0	1.0-2.0	2.5-3.5
Weanling foal, 6 months	0.5-1.0	1.5-3.0	2.0-3.5
Yearling foal, 12 months	1.0-1.5	1.0-2.0	2.0-3.0
Long yearling, 18 months	1.0-1.5	1.0-1.5	2.0-2.5
Two year old (24 months)	1.0-1.5	1.0-1.5	1.75-2.5

[a] air-dry feed (above 90% DM)

Source: Adapted from NRC (1989).

The energy requirements of mature horses at maintenance are low and can be met by feeding good-quality roughages (Table 3). The only supplemental feed required will probably be salt and a balanced mineral supplement provided free-choice. Because it is difficult for a horse to consume sufficient roughage to meet the energy needed for hard work, concentrates are normally required to meet working horses' energy needs. Thus, the concentrate-to-roughage ratio in diets for working horses should increase as the intensity or amount of work increases.

Table 3 : Digestible energy requirements for growth of foals

Age	Requirement (Mcal DE/kg of gain)
Weanling, 4 months	9.1
Weanling, 6 months	11.0
Yearling	15.5
Long yearling	18.4
2 years old	19.6

Source: Adapted from NRC (1989).

The quality of the water and feeds supplied to horses is very important. In many cases the water available to horses can have adverse or toxic effects. A high salt content and trace minerals contaminated water are undesirable, or in some cases unfit, for consumption. The NRC (1974) has published a list of the recommended upper limits for some potentially toxic substances in the drinking water of horses (Table 4).

Table 4 : Safe level of some mineral elements in drinking water for horses

Mineral Element	Safe Upper Limit (mg/liter)
Arsenic	0.2
Cadmium	0.05
Chromium	1.0
Cobalt	1.0
Copper	0.5
Fluoride	2.0
Lead	0.1
Mercury	0.01
Nickel	1.0
Nitrate nitrogen	100.0
Nitrite nitrogen	10.0
Vanadium	0.1
Zinc	25.0

Source: Adapted from NRC (1974).

6. Conclusion

Currently, DIHAR is studying hematology and blood biochemicals, high altitude (HA) stress related biomarkers, and work performance in pure breed as well as crossbred ponies. Reproductive performance, genotyping, nutrient requirements studies and development of ration scale are other R&D activities at DIHAR. Already this institute has issued ponies to the local populace as well as to the army for its logistic support. Recently 32 ponies of breeding stock have been issued to RVC for its further production and issue to the army in this region. Over all these animals are not only better adaptable to this region but also economically more viable to be used by the troops in high altitude regions.

References

National Research Council. (1974). Nutrients and toxic substances in water for livestock and poultry. National Academy Press, Washington, D.C.

Nutrient Requirements of Horses. (1989). 5th Edn. Committee on Animal Nutrition, National Research Council (NRC). National Academy Press, Washington, D.C.

National Research Council. (1980). Mineral tolerance of domestic animals. National Academy Press, Washington, D.C.

□□□

Chapter-18

Prospects of Fodder Production and Management in Cold Deserts

Janifer Raj X., Jitendra Kumar, Eli Paljor, Narendra Singh, O.P. Chaurasia and R.B. Srivastava
Defence Institute of High Altitude Research (DIHAR), DRDO, Leh, Ladakh

1. Introduction

The harsh climatic conditions prevailing in Ladakh results in very sparse vegetation. The cultivation of crops is limited to summers due to prevalence of sub zero temperatures for more than 6 months. Hence, livestock rearing plays a significant role in the economy of the Himalayan people, the forage plants play an important role in sustenance and development of livestock industry. The animal population of Leh is about 3,45,000 domestic sheep and goats, 58,000 cattle (including yak-cow hybrids), 5,500 horses, 7,500 donkeys, and 50 camels (Darokhan, 1986), which is about ten times of the human population of Ladakh (Joseph *et al.*, 1994). They have to survive and sustain on wild fodder plants and Lucerne which is the main crop cultivated for fodder purposes. Around 10 percent of the total geographical area, is used for the cultivation of cereals, pulses and oil seeds. Forests are the major sources of forage in this region. The very limited area of cultivation, coupled with average livestock productivity in Leh district reflects the very marginal quality of grazing and agricultural lands. Due to ever-increasing demand of animal products, the livestock population has increased manifold in recent post. As cultivation is possible only during summers people depend upon hay, mostly of Lucerne (*Medicago* sp.), as winter fodder. This chapter describes the present status of forage production, contributions of (DIHAR) and future strategies for the development of forage resource to ensure sustainable livestock production in Ladakh region.

2. Livestock Rearing in Ladakh

Grasslands are the major feed resource for livestock rearing in Ladakh. Climatic, topography, physiographic factors, altitude and related aspects have influenced the distribution of various grass species, which determine the grassland production qualitatively and quantitatively (Whyte, 1968). Though livestock rearing is an important occupation of farmers in this area, the forage cultivation has remained almost neglected. Grazing in the forest areas and sub-alpine and alpine pastures is the mainstay for the animals. Fodder trees and shrubs also contribute significantly. There is acute shortage of fodder especially green nutritious fodder due to climatic constraints and lack of scientific information for production of the fodder crops, which is the major cause of low productivity of the livestock (Roy, 1993). The shortage of fodder supply resulting in critically low biomass availability in the Himalayan region is mainly attributed to less area under fodder crops, uncontrolled grazing, poor management practices, heavy livestock population and over exploitation and unscientific management of fodder trees and pastures (Khosla and Toky, 1985). Consequently the livestock productivity is very low and all the Himalayan states have to import various livestock products from the plains.

3. Fodder Biodiversity of Ladakh

Ladakh region is dominated by flora of alpine and high alpine nature due to high altitude conditions. The lower alpine zones are covered with meadows, pasturelands, moraines and marshes. The higher reaches, which are covered with snow for more than six months in a year and have a bright sunshine (with more quantity of UV rays) are suitable for vegetation during the brief summer (Polunin and Stainton, 1989). The alpine species, e.g., sedges, grasses, fortes and shrubs are highly palatable and not too resistant to over grazing. The Himalayan grasslands experience intense grazing pressure on account of being the prime source of forage which can lead to the disappearance of nutritive species and infestation by less palatable species and weeds.

Table 1 : Nutritional Profile of Wild Fodder Plants of Ladakh

S. No	Botanical Name	Part Used	Moisture %	Dry Matter %	Crude Protein %	Neutral detergent fibre %	Acid detergent fibre %	Lignin %	Hemi cellulose %	Cellulose %
1.	*Aconogonum tortuosum*	Whole plant	-	88.87	8.63	51.04	38.22	12.63	12.82	25.59
2.	*Avena fatua*	Whole plant	09.09	90.91	6.79	38.35	28.96	04.60	09.39	24.34
3.	*Capparis spinosa*	Leaves	09.20	90.80	17.87	06.83	-	-	-	-
4.	*Chenopodium botrys*	Leaves	07.10	92.90	26.40	07.10	-	-	-	-
5.	*Cicer microphyllum*	Whole plant	08.60	91.40	15.30	36.38	32.34	05.90	04.04	26.44
6.	*Ephedra gerardiana*	Leaves & Fruits	08.19	91.81	09.27	45.23	42.48	21.28	02.75	21.20
7.	*Fagopyrum esculentum*	Whole plant	05.70	94.30	25.19	07.70	37.28	7.15	13.26	30.13
8.	*Fagopyrum tartaricum*	Whole plant	03.22	96.78	20.12	06.80	-	-	-	-
9.	*Hippophae rhamnoides*	Leaves	08.99	91.09	23.93	22.72	20.21	08.47	02.51	11.74
10.	*Iris lactea*	Leaves	09.63	90.37	10.62	47.28	42.35	04.01	-	-
11.	*Lepidium latifolium*	Whole plant	08.26	91.74	24.62	11.20	-	-	-	-

12.	*Medicago falcata*	Whole plant	-	-	15.85	20.86	-	-	-	25.80
13.	*Medicago media*	Whole plant	-	-	17.50	20.58	-	-	-	25.70
14.	*Medicago sativa*	Whole plant	-	-	18.22	20.65	-	-	-	25.90
15.	*Melilotus officinalis*	Whole plant	10.05	89.95	24.01	20.94	18.53	04.71	02.41	13.82
16.	*Prangos pabularia*	Whole plant	10.57	89.43	12.66	35.78	32.69	05.12	03.09	27.57
17.	*Rheum webbianum*	Leaves	08.70	91.30	12.80	14.95	-	-	-	-
18.	*Rumex spp*	Whole plant		88.38	23.57	30.89	21.00	7.29	9.89	13.71
19.	*Salix babylonica*	Leaves	08.97	91.03	18.12	36.35	24.68	09.28	11.67	15.40
20.	*Trifolium partense*	Whole plant	10.32	89.68	17.56	39.14	32.56	06.23	06.58	26.35
21.	*Trifolium repens*	Whole plant	10.53	89.47	12.60	41.80	32.75	06.87	09.05	25.88
22.	*Ulmus wallichiana*	Leaves		89.70	19.02	28.98	22.74	6.05	6.24	16.69

Important fodder crops of temperate region are; *Avena sativa, Brassica* sp., *Medicago sativa* and *Pisum sativum*. The major temperate grasses which offer significant diversity in these regions are: *Agrostis, Agropyron, Dactylis, Elymus, Festuca, Lolium, Phalaris, Phleum* and *Stipa*. The major temperate legumes are: *Astragalus, Lespodeza, Lotus, Medicago, Melilotus* and *Trifolium*. Forage legumes are important because they enrich the nitrogen content of the soil and have high nutritive value and can be grown in mixtures with grasses in grasslands. Among the legumes, Siratro (*Macroptelium atropurpureum*), *Stylosanthes hamata, S. scabra, Glycine javanica, Dolichos auxilaris, Desmodium* sp. and *Centrosema pubescens* show good performance (Melkania, 1995).

In the case of forage legumes the diversity is extensive and the classical example of natural variations is that of *Medicago* (Fig. 1). Ladakh and adjoining areas are considered as one of the centers of origin for *Medicago* and *M. falcata* is the original species found here. The traders who used to travel through Ladakh along the old Silk Route noticed better forage yield of *M. sativa* in Yarkand (Uzbekistan) and introduced it in Ladakh. Now both *M. falcata* and *M. sativa* have hybridized to such an extent that at least five undifferentiated forms are found (Misri, 1986). The variability in the habit of plants is significant. In a single field erect, decumbent or prostrate types of plants can be found. The variability also occurs in the height, leaf size, branch length and their number. Pod shape is one of the diagnostic features for identification of various species of *Medicago*. The pod of *Medicago sativa* is double spiral or much curved, small, indehiscent, smooth or spiny; and one to few seeded (Bailey, 1968). In the case of *M. falcata* the pod is linear, sickle shaped, glabrous with 5-10 seeds (Hooker, 1897). The present populations of Ladakh *Medicago* are assigned to the variegated group of *Medicago* classification proposed by Whyte (1953). Newly reclaimed areas for agriculture are sown under Lucerne for 2-3 years before crop production is initiated. The month of May is considered optimum for sowing of seeds in rows at a depth of 4-5 cm (15 kg/ha) as irrigation becomes available on melting of snow. Most of the Lucerne fields are 10-20 years old and they have not been re-sown due to self-seeding.

Lucerne is also grown as intercrop in orchards and two cuts are harvested during the season. Upto mid September two cuttings are taken and stored as winter feed for animals. Dry matter yield of Lucerne was 90.76 percent DM in pellet form. Green forage is reported to contain 80.0 percent moisture, 5.2g protein, 0.9g fat, 3.5g fibre and

2.4g ash per 100g. Lucerne whole meal and leaf meal are reported to contain 66 and 77 calories, 7.5 and 8.0 percent moisture, 16.0 and 20.4g protein, 2.5 and 2.6g fat, 27.3 and 17.1g fibre, 9.1 and 11.5g ash per 100g, respectively. Lucerne is a valuable source of vitamins A and E and contains β-carotene 6.24, thiamine 0.15, riboflavin 0.46, niacin 1.81,tocopherol 0.46 and niacin 1.81 while pantothenic acid, biotin, folic acid, choline, inositol, pyridoxine, vitamin B_{12} and vitamin K are also present. Fresh Lucerne is rich in vitamin C (1.78 mg/g) but it loses 80 percent on drying (Duke, 1983).

Medicago sativa *M. falcata*

M. media

Fig. 1 : *Medicago* species of Ladakh

4. Present Status of Forage Production in Ladakh

The total area of Ladakh is 95,876 km^2 of which 28 percent is cultivable (this area is confined to lower valleys where irrigation from adjoining streams and rivers is available). The principle vegetative growth starts with the beginning of summer in April when the melting of the snow provides moisture to the herbaceous vegetation. The flora is in full bloom by August and starts to degenerate by the end of September (Kachroo *et al.*, 1977). Forage cultivation is restricted to only about one per cent of the cultivated area in the entire Himalayan region because of the preponderance of marginal and small land holdings in the area. Lucerne is cultivated in an area of 4,000 ha where assured irrigation is available. Misri (1988) studied the biomass availability of some of the representative pastures of Kashmir Himalaya and found that green herbage availability varied from 4.7 to 29.1 t/ha. The estimated net above ground biomass varied from 279 to 1568 gm^{-2} for low elevation Himalaya, 219 to 285 gm^{-2} for mid elevation Himalaya and 233-372 gm^{-2} for high elevation Himalaya (Melkania and Singh, 1989). Besides grazing and fodder trees, the major local forage resource is the crop residue, which again is too inadequate to sustain the livestock.

5. Contributions of DIHAR For Fodder Security in Ladakh

Keeping in view the constraints in fodder production and to overcome the gap between demand and supply emphasis is given by DIHAR on several steps for augmenting the fodder production in Ladakh. The strategies for improvement and conservation of Himalayan resources particularly the forage resources through the scientific interventions could help in improving the productivity of forages. As fodder production is a component of the farming system efforts are being made for increasing the forage production in a farming system approach. The forage productivity could be enhanced with the adoption of the mentioned technologies standardized by DIHAR such as:

5.1 Nutritional Analysis and Hay Making of Lucerne

As Lucerne is the important fodder crop of this region DIHAR has analyzed its nutritional value of Lucerne. All the three species of Lucerne found in Ladakh namely *Medicago sativa*, *M. falcata* and *M. media* are highly nutritive (Table 2).

Table 2 : Chemical composition (%) of different species of medicago on dry matter basis

Nutrient	*M.sativa*	*M.media*	*M.falcata*
Crude protein(g)	18.22	17.50	15.85
Crude fibre (g)	20.65	20.58	20.86
Total Ash(g)	5.14	5.64	7.26
Non- Nitrogenous Extracted Substances	54.42	55.05	54.89

DIHAR has also standardized the hay making for winter feed purposes minimizing nutritional losses. For high quality hay, the crop has to be cut at one tenth to one fourth at bloom stage or when its lower leaves start turning yellow and the maximum quality of nutrients are retained in hay at this stage of harvest. The harvest at bloom stage is also advised as it gives higher yield and protects the crop stand. The farmers bundle the crop after harvest and stack on the roofs of the houses but the traditional method leads to loss of more leaves and only the stem remains while feeding. Due to problems such as loss of leaves due to handling, loss of carotene and other nutrients due to bleaching on exposure to sun light, the hay can be baled and stored for further use. The green fodder is dried to reduce the moisture content to 25 percent and the drying is very rapid in Ladakh due to lesser atmospheric humidity. The bundles of fodder are allowed in the field for 2-3 sunny days and turned upside down daily to ensure proper drying. The process of curing follows drying which reduces the moisture content to 15-18 percent and ferments the hay for better aroma and palatability. DIHAR has standardized baling of Lucerne and bales of standard size (25 kg) are made using the baling press manufactured by HIMCO, Bareilly and the finely powdered hay collected during baling is an ideal feed for the poultry. Baling of Lucerne reduces leaf shed, easy transportation, minimal bleaching, less storage space and easy accounting of hay stocks (Chaurasia *et al.,* 2004).

5.2 Identification of Native Fodder Plants

DIHAR has collected and identified more than 1000 species of wild plants and many of them have been observed to be useful as fodder to sustain animal husbandry in Ladakh. These potential wild fodder plants are capable of growing in the barren mountains of Ladakh during summers. The fodder samples were also analyzed for the nutritional values in collaboration with Indian Grass and Forage Research Institute (IGFRI), Jhansi (Table 2).

5.3 Feed Formulation using Local Ingredients

Feed formulation including the local feed material such as seabuckthorn leaves, apricot cake could be useful in improving the productivity of the livestock. DIHAR has studied the effect of seabuckthorn leaves and Lucerne hay based complete diet on growth and blood chemical profile of sheep at high altitudes and found that besides the inferior effect of feeding seabuckthorn leaves compared to Lucerne hay, cost benefit analysis revealed that seabuckthorn leaves can be used economically as a local feed ingredient in feeding sheep (Jadhav *et al.,* 2009).

5.4 Introduction of Legumes / Grasses / Fodder Tree

Introduction of improved grass/legume species in the native vegetation not only helps in increasing total biomass production but also improve the quantity of the produce. DIHAR is involved in introducing fodder crops in Ladakh region. Foliage of fodder trees such as *Ulmus, Salix* sp. could be fed to the livestock in mixture with crop residues and hay as mixing of tree foliage with dry roughage improves the palatability and nutritive value. The herbage yield was found five times more and nutritive value was found two times higher than the local hay when compared with grasses-legume mixtures. DIHAR has introduced and standardized the cultivation practices for *Brassica* sp., *Medicago sativa* and *Pisum sativum* which could serve as better fodder species for Ladakh, to be increasingly cultivated along with traditional fodder. DIHAR has standardized the cultivation techniques for oil seed crops such as Mustard and Sunflower. The oil seed crop cultivation could be taken up by the farmers and oil seed concentrate obtained after oil extraction for culinary purposes could be used as feed for animal as concentrate to increase their milk yield and body weight.

Recently, *Onobrychis vicifolia* commonly called as Sainfoin is being evaluated for introduction (Fig 2). It is a pubescent cross-pollinated perennial with many erect or sub-erect, hollow stems, 60–80 cm or more, arising from basal buds on a branched root stock. This forage does not cause bloat in animals and is highly acceptable, leading to high animal intake of forage.

Figure 2 : *Onobrychis viciifolia*

5.5 Genetic Characterization of *Medicago* sp.

The native biodiversity of fodders is huge and their conservation for future breeding programmes to develop cultivars suitable for this region is given importance. Genetic diversity of Twenty five accessions of Lucerne (*Medicago* sp.) collected from Leh valley of Trans-Himalayan region of Ladakh, Jammu and Kashmir was analyzed using Inter Simple Sequence Repeats (ISSRs) and Random Amplified Polymorphic DNA (RAPD). Our results revealed that the level of genetic variation in the *Medicago* ecotypes was relatively high. The mean coefficient of differentiation (Gst) was 0.0584 indicating 30.23 percent of the genetic diversity within the populations. The overall value of mean estimated number of gene flow (Nm = 8.0682) revealed large gene exchanges among populations and attributed to the cross pollination. The plant is capable of reproducing by self-sowing, thus can influence population genetic structure. The pronounced genetic variation indicates that *Medicago* species is a proper plant for genetic research and that there is great potential of using this plant in further breeding programmes to develop cultivars suitable for this region.

6. Future Prospects

The information gathered reveals that Lucerne is the only cultivated fodder crop. As such there is an urgent need to enhance fodder availability in Ladakh. With the increase in the domestic animal population in the area and the necessity of higher animal product yields, there has been a progressive increase for both quantity and quality nutritive fodder. Silage friendly plants such as *Salix* sp. which are grown during summers can be preserved as silage to be used during lean periods when green fodder is scarce especially during winters. Participatory techniques should be adopted to carry out fodder improvement programme. The establishment of community silage pit in a village will be beneficial as there will be continuous supply of green fodder for the whole village during winters. Fodder tree improvement programmes for higher leaf fodder should be initiated. Fodder production can also be improved by growing forage grasses and fodder trees along village roads and panchayat lands. The present poor production potential of pastures could also be attributed to poor fertility of soils. To raise the fertility status and rectify the deficiencies, soil testing coupled with field trials should be conducted to work out the fertilizer requirement of different pastures. In short, integration of forage crops in the existing crop sequences, growing food cum fodder crops, intensive forage cultivation, developing suitable cropping systems in addition to natural pasture and wasteland improvement are some of the integrated approaches for solving the forage problem of the cold desert.

References

Bailey, L.H. (1968). Manual of cultivated Plants. The Macmillan Co., New York.

Chaurasia, O.P., Paljor, E., Ballabh, B., Prakash, S., Ahmed, Z. and Raut, B. (2004). Alfalfa (Ole) Production in Ladakh .*Extension folder No.10*.Feild research Laboratory, DRDO, Leh.

Darokhan, M.D. (1986). Animal husbandry in Ladakh: an ecological perspective. Ladakh Project, *Ecology and Principles for Sustainable Development*, Proceedings of a conference co-hosted by the Ladakh Project and the Ladakh Ecological Development Group in Leh,Ladakh, pp. 71-74.

Duke, J.A. (1983). Handbook of Energy crops. http: hort.purdue.edu/newcrop/ duke-energy /Medicago-sativa. Html#cultivation.

Hooker, J.D. (1897). Flora of British India vol II. L.Reeves & Co., London.

Jadhav, S., Tilak Raj, Guru Charan and Singh, S. B. (2009). Effect of Seabuckthorn Leaves and Lucerne Hay Based Complete Diets on Growth And Blood Chemical Profile Of Sheep At High Altitudes. In:Advances in Agriculture environment and health.Eds.SB Singh, OP Chaursia, Ashish Yadav, AM Rimando and TH Terrill. Sathish Serial Publishing Company, NewDelhi.pp.415-423.

Joseph, L., Fox, C. N., Bhatt, S. and Chandola, A. (1994).Wildlife Conservation And Land-Use Changes in the Trans Himalayan Region of Ladakh, India. *Mountain Research and Development*, 14 (1) :39-60.

Kachroo, P., Sapru, B. and Dhar, U. (1977). *Flora of Ladakh, an Ecological and Taxonomical Appraisal* Risen Singh and Mahendra Pal Singh Publ., New Delhi.

Khosla, P.K. and Toky, O. P. (1985). Renewed scientific interest in agroforesry. In: Agroforestry systems: A New Challenge (ed. P.K. Khosla and S. Puri) ISTS, Solan : 7-24.

Melkania, N.P. and Singh, J.S (1989). Ecology of Indian grasslands. In: Perspective in Ecology (ed. J.S Singh and B. Gopal), Jagmonder Book Agency, New Delhi.

Melkania, N.P. (1995). Forage resources and forage production systems in Central Himalaya. In: New Vistas in Forage production (ed. Hazra, C.R and Misri, Bimal). AICRPF (IGFRI). Publication Information Directorate, New Delhi. 203-211.

Misri, B.(1986).Variability in Alfalfa of Ladakh. Plant Germplasm Collection Report http://www.Fao.org/Ag/Agp/Agpc/Doc/Bulletin/Ladakh. Htm.

Misri, B.K. (1988). Forage production in alpine and sub-alpine regions on northwestern Himalaya. In: Pasture and Forage Crop Research, A State of Knowledge Report (ed. Punjab Singh). RMSI, Jhansi. 43-45.

Polunin, O. and Stainton, A. (1989). Flowers of the Himalaya. Oxford Univ. Press, New Delhi.

Roy, D. (1993). Reap more biomass through diversity in forestry. *Intensive Agriculture*, 31 : 23-26.

Whyte, R.O. (1968). Grasslands of Monsoon. Faber and Faber, London. pp. 325

Whyte, R.O., Leissner N. and Trumble H.C. (1953). Legumes in Agriculture. FAO, Rome.

❑❑❑

Chapter-19

Eco-Environment Improvement of Cold Desert : *DIHAR Initiatives*

Pal Murugan M., Eli Paljor, Janifer Raj X., O.P. Chaurasia, Narendra Singh and R.B. Srivastava
Defence Institute of High Altitude Research (DIHAR), DRDO, Leh, Ladakh

1. Introduction

Owing to prolonged subzero temperature and non-availability of water, cultivation is only possible during the summer season between March-September in the Trans-Himalayas Zone. The region has very meager vegetation with less number of tree species which necessitated Defence Institute if High Altitude Research (DIHAR) to take key initiatives to make the inaccessible and hostile terrain of Ladakh reasonably green through its various research and development activities. The high altitude, low vegetation and partial air pressure causes hypoxia which results in poor work efficiency and other psycho physiological problems amidst the local population and troops deployed in this high altitude region. Extensive plantation of perennial trees, shrubs, cultivation of vegetables, fruits and medicinal plants are considered to be the most amicable solutions to these problems. However, survival of young saplings is poor (55% to 65%) because of the harsh atmospheric conditions and short growing period (5 to 6 months). The harsh atmospheric condition, short growing season, the subzero temperature and land locked snow bound condition gave lot of hurdles to the scientists and working staff of DIHAR in developing cold arid agro techniques for tree plantation in this region. In spite of multitude of constraints DIHAR has developed various agro techniques in the field of forest trees, ornamental plants, agro forestry, floriculture and ornamental gardening to create green cover in Ladakh. This chapter gives a brief account about the efforts taken by DIHAR in collaboration with the State Forest Department, NGOs and Army Sadbhavana Cell towards the creation of greener Ladakh.

2. Research on Flora of Ladakh

The vegetation of Ladakh can be classified into four categories such as Alpine mesophytes, Oasiatic vegetation, Desert vegetation and High alpine vegetation dominated by annual and perennial herbs, followed by few stunted shrubs and bushes. DIHAR has carried out extensive surveys since 1993 and provided the tentative statistical analysis of the flora of Ladakh Himalayas (Chaurasia *et al.*, 2003; Table 1). The forest area accounts for approximately 200 km^2 including seabuckthorn in revenue and community land out of the total area of 45,113 km^2. DIHAR has also identified and documented 41 trees and shrub species for increasing the forest cover (Chaurasia *et al.*, 2007; Table 2).

Table 1 : Statistical analysis of Flora of Ladakh

Group	Families	Genera	Species
Angiosperm	60	342	698
Dicoteyledon	54	326	637
Monocotyledon	06	16	61
Gymnosperm	02	02	04

Table 2 : List of forest trees and shrubs of Ladakh

Name of the plant	Family	Uses
Caragana vesicolor	Fabaceae	Fuel purpose
Colutea multiflora	Fabaceae	Potential forest shrub
Cotoneaster falconeri	Rosaceae	Potential forest shrub
Crataegus songarica	Rosaceae	Timber purposes
Elaegnus angustifolia	Elagnaceae	Ornamental purposes

Fraxinus xanthoxyloides	Oleaceae	Forest tree
Hippophae rhamnoides	Elagnaceae	Forest shrub
Hippophae tibetana	Elagnaceae	High alpine forestry
Hippophae salicifolia	Elagnaceae	Forest shrub
Indigofera heterantha	Fabaceae	Forest plant
Juglans regia	Jugulandaceae	Timber purposes
Juniperus communis	Cupressaceac	Ornamental and religious purposes
Juniperus macropoda	Cupressaceae	Ornamental and religious purposes
Juniperus recurva	Cupressaceae	Ornamental and religious purposes
Morus alba	Moraceae	Fruit is edible
Myricaria elegans	Tamaricaceae	Forest shrub
Myricaria germanica	Tamaricaceae	Forest shrub
Myricaria squamosa	Tamaricaceae	Forest shrub
Picea smithiana	Pinaceae	Timber purposes
Populus caspica	Saliaceae	Valuable timber tree
Populus ciliata	Saliaceae	Valuable timber tree
Populus deltoides	Saliaceae	Valuable timber tree
Populus euphratica	Saliaceae	Valuable timber tree
Populus nigra	Saliaceae	Valuable timber tree
Populus sylvestris	Saliaceae	Valuable timber tree
Robinia pseudacacia	Fabaceae	Forest tree
Rosa webbiana	Rosaceae	Ornamental shrub
Salix alba	Saliaceae	Forest tree
Salix angustifolia	Saliaceae	Forest tree
Salix babylonica	Saliaceae	Ornamental tree
Salix caesia	Saliaceae	Forest tree
Salix daphnoides	Saliaceae	Forest tree
Salix denticulata	Saliaceae	Forest tree
Salix excelsa	Saliaceae	Forest tree
Salix hastata	Saliaceae	Forest tree
Salix sclerophylla	Saliaceae	Forest tree
Salix tetrasperma	Saliaceae	Forest tree
Sophora moorcroftiana	Fabaceae	Forest shrub
Syringa emodi	Oleaceae	Ornamental tree
Tamarix gallica	Tamaricaceae	Forest shrub
Ulmus wallichiana	Ulmaceae	Forest tree

3. Phenological Observations on Trees of Ladakh

Phenology is the study of relationship between climatic factors and periodic phenomena in organisms. The word phenology was derived from the greek word *Phaino-* meaning to show or to appear. The alpine flora of Ladakh exhibits numerous ecological, physiological, morphological and reproductive adaptations to withstand the harsh climatic condition. The tree species and some thorny bushes are mostly seen in the valleys except Juniper which is available in altitudes of more than 10000 ft. Willow, Poplar, Apple, Apricot, Walnut, *Robinia,* Wild rose, Ulmus, Oleaster, Seabuckthorn, Mulberry and *Myricaria* are some of the dominant woody species available widely in the five valleys of Ladakh. DIHAR has carried out the phenological observations on these tree species during 2008-2009 as study of plant phenology provides knowledge about the pattern of plant growth and development as well as the effects of environment and selective pressures on flowering and fruiting behavior (Zhang *et al.,* 2006). The phenological data could also reveal the potential impacts of future climate changes (Karmer, 1997). The results of phenological observations indicated that trees such as Apricot (*Prunus armeniaca),* Poplar (*Populus caspica),* Willow (*Salix denticulate)* exhibit flowering phase even before the new foliage phase. The fruiting and leaf flushing occurs concurrently or later. In the trees like Walnut (*Juglans regia*), Wild Rose (*Rosa webbiana*), Seabuckthorn (*Hippophae rhamnoides*), Mulberry (*Morus alba*), Ulmus (*Ulmus wallichiana*) and Oleaster (*Elaeagnus angustifolia*) leaf flushing occurs first followed by reproductive phase. It was also observed that the raise in temperature from subzero to above zero during the last week of March and April months favours the new foliage in most of the tree species except in apple, apricot, ulmus and seabuckthorn.

Primack (1985) reported that species with larger fruits (requiring a longer time for fruit development) flowered earlier than species with small fruits. In the present study also trees with larger fruits such as Apple and Apricot, flowering occurred on bare branches and leaf flushing occurred later while in Seabuckthorn which requires lesser time for fruit development the reproductive phase followed the leaf flushing phase. Leaf-fall initiation was a periodic activity in all the deciduous species studied. However, the onset of leaf fall initiation is different in various woody species. In most plant species, leaf shedding begins in the month of October with peak in November - December with the decrease in temperature. By the end of December all the woody species studied are devoid of leaves and by the end of December all the woody species studied were devoid of leaves (Murugan *et al.*, 2010).

4. Carbon sequestration in Nubra Valley

Fruit tree orchards have received attention in recent years by virtue of their potential to act as carbon pools. The Nubra Valley at an altitude of 3,048 m, is covered by > 60,000 temperate fruit trees. A study was conducted during 2008-2009 by DIHAR on the potential of these fruit-yielding trees as means of carbon sequestration in the Nubra Valley as the valley produces approx. 4,650 metric tonnes (MT) of apricot, apple, walnut, mulberry, grape, and plum fruit. It was observed that the total biomass of the temperate fruit trees (i.e., apricot, apple, and walnut) in the area surveyed was 40,000 MT dry weight (DW). Thus, 22,300 MT of carbon has been sequestered by these fruit species. The study revealed that the cultivation of temperate fruit crops species such as apricot, apple, and walnut could be exploited more efficiently for carbon sequestration which subsequently helps in mitigating global CO_2 levels and restoring degraded land in these cold arid areas by improving soil organic carbon (SOC) contents. These results concluded, in part, that growing biomass to sequester carbon is a beneficial grow-and-store process for sustainable development.

It is recommended from the study that the villagers in the Nubra Valley should be encouraged to compost fruit tree leaf litter using composting pits along with adding sustainable micro-/macro-fauna and composting bacteria. The resulting compost could then be used in their orchards to increase both the soil fertility status and carbon sequestration. The potential for gaining carbon (CER) credits based on the total biomass of the fruit orchards in the Nubra Valley is limited compared to the major national forest stands. This type of carbon sequestration programme will encourage villagers to become more involved in global carbon-emissions mitigation and to improve their socio-economy through increased orchard production. This practice appears to offer a relatively low-cost approach to enhance the sequestering of carbon. Furthermore, orchard technology seems to be well-developed and effective in the Nubra Valley. So far, few villages have planted orchards and many are unaware of their potential value. The study suggests that agricultural organizations should collaborate and encourage orchard plantations in all villages of the Nubra Valley (Kumar *et al.*, 2010).

5. Green Drive of DIHAR towards Greener Ladakh

5.1 Inputs Supplied by DIHAR for Greening of Ladakh

5.1.1 Nursery and mother plants of trees

DIHAR has developed more than 200 mother plants of Willow and more than 100 mother plants of Poplar from which the cuttings are made through its agro forestry division. It also has high quality nursery of Apricot and Seabuckthorn. The fast growing high quality Seabuckthorn plant variety developed by DIHAR, which grows more than 6 feet height is also used for the green drive in Ladakh. The fruit trees such as Apricot (*Prunus armeniaca*), Apple (*Malus domestica*) and Walnut (*Juglans regia*) do well in this region without much care. The saplings of Apple and Walnut are also supplied by Horticulture division of DIHAR to the army and local population. Every year DIHAR is supplying more than 20,000 saplings of Willow, Poplar Seabuckthorn and Apricot to the army and local community for greening of Ladakh (Table 3).

Table 3 : Distribution of Tree and Fruit saplings to Army Units and Farmers for afforestation from 2006-2011

Year	Total	Willow	Poplar	Apricot	Apple	Others (Walnut, Peach, Plum & Cherry)
2006	63478	39792	9686	14000	-	-
2007	33101	25067	5350	2684	-	-
2008	15069	9060	3215	2794	-	-
2009	19000	12000	5000	2000	-	-
2010	15500	11500	3000	1000	-	-
2011	27600	14000	5000	600	6000	2000
Total	1,38,448	87219	23151	20078	6000	2000

5.1.2 Seeds and saplings of vegetables

Before the inception of DIHAR, the people of Ladakh only cultivated wheat, barley, potato and local turnip for their consumption. By intensive research trials, DIHAR has identified 78 types of suitable vegetables and their improved varieties for cultivation in the Ladakh region. By adopting these technologies, local farmers are able to supply 50 percent of the vegetable requirement of the troops in this region. Remoteness and technological backwardness of this area hinders the availability of quality seeds and seedling for the cultivation. As stated earlier, the harsh climatic condition prevailing in Ladakh does not favour early supply of vegetable seedlings for cultivation. To overcome these problems and to boost the fresh food availability and to create aesthetic floral garden DIHAR supplies more than 12 Lakh seedlings of vegetable and flowers every year during month of May (Table 4).

Table 4 : Year wise Supply of seedlings and saplings of vegetables and ornamentals from DIHAR, Leh and DIHAR Det, Paratpur from 2006-2011.

Year	Amount of seedlings and saplings distributed (in numbers)
2006	9,80,500
2007	11,50,000
2008	9,00,000
2009	12,50,000
2010	12,50,000
2011	13,00,000

5.1.3 Organic manure for soil health

Towards the sustainability of afforestation programme in Ladakh DIHAR is providing various inputs such as manures and fertilizers to the farming community and troops deployed in the far flung areas of Ladakh. The Farm Yard Manure (FYM) accumulated in the dairy, poultry and equine farms of DIHAR and its Det Partapur is also supplied to the needy people to enhance the soil fertility of the least fertile virgin soils of Ladakh. The addition of organic matter in the form of farm yard manure not only enhances the soil fertility but also improves the water holding capacity, microbial activity, soil texture and structure. By visualizing the benefits of organic manure application to the soil many army units located in the forward posts and the civil population demand more farm yard manure to enhance the fertility of the soil.

5.2 Technical Support Provided by DIHAR

5.2.1 Training the locals and troops

Every year DIHAR is organizing various trainings during summer and winter for the army personnel and the local farming community on vegetable cultivation, arboriculture, fruit and flower cultivation to green this region. Syllabus of training mainly covers the following aspects:

a) Protected vegetable nursery/tree sapling raising techniques.
b) Protected vegetable cultivation techniques.

c) Techniques of *in-situ* seed production of certain vegetables.
d) Techniques of tree plantation and plantation management.
e) Techniques of fruit and flower cultivation.

Besides in campus trainings, DIHAR is also providing more than 100 on the spot training-cum-demonstrations every year on preparation of layout plan, beds, pits, seed sowing, tree planting, intercultural operations and pest and disease management practices for the troops and local people. On farm demonstration is also conducted in farmers' fields to popularize the latest agro-animal technologies in different villages of Ladakh.

5.2.2 Field management techniques

DIHAR, over the five decades through its various research and development projects, has identified the tree and shrub species suitable for this region, standardized the propagation techniques for massive plantation in this undulated terrain of Ladakh and to provide impetus to the afforestation programme. DIHAR has published an extension folder titled '*Planting Tips for a Forester*' in the year 2001. Dwivedi *et al.* (2001) reported the technical guidelines for successful implementation of the afforestation programme.

a. Selection of Plant Species

Willow (*Salix regia*) and Poplar (*Populus indica*) are widely adopted tree species because of its economic value as timber and fuel. Robinia (*Rubinia sp*), a leguminous thorny forest tree is another potential plant which can be used for plantation up to an altitude of 10,000 ft above msl. Naturally growing plants/shrubs like Elm (*Ulmus walghiana*), Tamarix, Oleaster (*Elaegnous hortensis*) Seabuckthorn (*Hippophae ramnoides*) and Wild Rose (*Rosa webbiana*) can be used for planned plantation since they have the natural ability to survive the climatic conditions particular to high altitude region.

b. Selection of Site and Land Preparation

DIHAR has also identified the field conditions suitable for growing trees in Ladakh. As soil depth is crucial for the survival of tree species the soil of the plantation area should be at least 3 feet deep and should be free from rocks and stones. Fertile sites in sunny locations with an assured source of irrigation should be selected. Since the soils of Ladakh are poor in organic matter and other nutrients and also lack the requisite depth, it becomes necessary to prepare pits for planting of the trees carefully. Before the onset of winter, pits having diameter and depth of 2-3 ft have to be dug and should be filled with a soil mixture comprising of farmyard manure, fertile soil and sand (2: 1: 1). A handful of DAP should be mixed well with this soil mixture. The pit is then irrigated well so that the soil mixture settles down.

c. Method of Propagation and Time of Planting

Forest trees are generally propagated through seeds and stem cuttings and in different spacings (Table 5). The seedlings are able to establish well developed root system. The stem cuttings should be planted with more care as those should be taken from healthy plants and planted in the pits and watered regularly. To fasten the root establishment and growth of the young saplings, DIHAR has standardized Plant Growth Regulator treatments with IAA (Indole Acetic Acid), NAA (Naphthalene Acetic Acid) and IBA (Indole Butyric Acid). The survival percentage and root establishment has increased considerably up to 80 percent by the IAA (Indole Acetic Acid) treatment for Seabuckthorn and Poplar cuttings. The planting should be done before the sprouting of the dormant shoots of winter and generally between mid March to mid April as the sprouted cutting did not establish well.

Table 5 : Method of propagation and planting distances for tree species

Trees	Method of propagation	Planting Distance (ft)
Poplar	3-5 ft long stem cuttings	8-10
Willow	3-5 ft long stem cuttings	10-12
Robinia	Seed	10-12
Elm	1 ft long stem cuttings	8-10
Seabuckthorn	1 ft long stem cuttings	3-4
Wild Rose	Suckers	3-4
Apricot	Seed	8-10
Walnut	Seed	12-14
Apple	Seed	8-10

d. Field Management

Frequent and light 2-3 irrigations per week after planting are essential. Mulching also helps to preserve soil moisture and increases soil temperature which helps in better establishment. Pruning from second year helps these to attain desired shape. When trees become dormant during winters no irrigation or fertilizer is to be applied. On sprouting normal care is to be given to the trees to maintain their vigour and growth.

6. Benefits of Greener Ladakh

The massive afforestation programme will result in the increased availability of oxygen in the oxygen deficient zone and give protection from intense UV radiation. It also reduces the intensity of dust storms frequent in Ladakh and thereby decrease the wind erosion. It is helpful in improving the ecology of this region by increasing the fertility of virgin soils of Ladakh and conserves available moisture in this region. The cultivation of tree species and shrubs aids in more timber and firewood availability for the local community. The vegetative cover created with various shrubs and trees also accommodate fauna thereby providing them shelter and food. The creation of aesthetic recreational ecology in army units reduces the psycho-physiological problems faced by army and Para military forces deployed in this snow bound terrain. The efforts taken by DIHAR along with army units and civil administration has resulted in more availability of fresh food in the cold arid region and make the people of Ladakh self sufficient in their food requirements and enhance their standard of living.

7. Recognitions

DIHAR was awarded the 'Indira Priyadarshini Award' by the Ministry of Environment and Forest, Government of India, in the year 1993.

"SAVE THE NATURE, PAVE THE FUTURE"

References

Chaurasia, O.P., Ballabh, B. and Raut, B. (2003). Plant Diversity of Ladakh Himalayas. FRL Extension folder No. 5, Leh, Ladakh.

Chaurasia, O.P., Ahmed, Z. and Ballabh, B. (2007). Ethnobotany and plants of Trans Himalayas. Satish Serial Publishing House, New Delhi.

Dwivedi, S.K., Paljor, E. and Attrey, D.P. (2001). Planting Tips for A Forester. FRL Extension folder No. 4, Leh, Ladakh.

Karmer, K. (1997). Phenology and growth of European trees in relation to climate change. pp. 39-50. In: H. Leith & M.D. Schwartz (ed.) Phenology of Seasonal Climates I. Bakhuys, The Netherlands.

Kumar, G.P., Murugan, M.P., Murkute, A.A. and Singh., S.B. (2010). Carbon Sequestration strategy by Temperate Fruit crops in Trans- Himalyan Region. *Journal of Horticultural Science and Biotechnology*.85 (5): 405-409.

Murugan, M.P., Janifer Raj. X, Kumar G. P., Gupta, S., Paljor E. and Singh., S.B. (2010). Phenological observations on temperate dry deciduous trees of the cold arid region Ladakh, Jammu and Kashmir. *Journal of Economic and Taxonomic Botany*.34 (4):833-839.

Primack, R.B. (1985). Patterns of flowering phenology in communities, populations, individuals and single flowers. In: White J, ed. The population structure of vegetation. Dordrecht: Junk, 571–593.

Zhang, G., Song Q. and Yang, D. (2006). Phenology of *Ficus racemosa* in Xishuangbanna, Southwest China . *Biotropica*. 38: 334-341.

□□□

Chapter-20

Potential of Non Conventional Energy Sources in Himalayan Frontiers

Syed B. Mustafa, Somen Acharya, Banu Prakash and R.B. Srivastava
Defence Institute of High Altitude Research (DIHAR), DRDO, Leh, Ladakh

1. Introduction

The survival and prosperity of all civilizations ability to access energy in sufficient quantities in the form of food and fuel has been the key. To meet the future energy demand and to give quality and pollution free supply to the growing population, the world to day is focusing on natural, clean, renewable and non conventional energy sources. These energy sources capture their energy from on-going natural processes such as geothermal heat flows, sunshine, wind, flowing water and biological processes.

Ladakh is the northern most part of India having an area of 92,000 sq. km, lying at an altitude of 8,500 to 15,000 ft. This Trans-Himalayan land is an icy barren landscape. Glistering white snow capped peaks and glaciers dominate in this most inhospitable terrain of its kind in the world. Ladakh is gifted with rich resources of non conventional energy like geothermal, wind and solar energy. There is a need to exploit all these resources properly for the development of the region and welfare of the people. Harnessing the non-conventional energy resources of Ladakh and its optimum utilization will provide clean energy to Ladakh region thereby reducing the usage of diesel and kerosene in the area.

2. Geothermal Energy Potential of Ladakh

India has been one of the earliest countries to begin geothermal projects way back in the 1970s, geothermal energy has however not emerged as a significant renewable energy option in the country. The preference, instead, is for solar energy and wind energy presumably on account of their easy availability, lower initial investment and proven technology. A projection of requirements two decades hence indicates that even at present levels of consumption, Indian annual import of oil would go up to 750 million tons and of coal to 1300 million tons.

Among the various new and renewable energy options, geothermal energy is recognized as the cleanest, without any environmental hazards. If a proper well at adequate temperature is identified, the energy source is limitless with the added advantage of minimal operating costs, unlike the case of wind or solar power. India has several geothermal provinces characterized by high heat flow and thermal gradients. Subsequent to the oil crisis of the 1970s, the Geological Survey of India generated data of resource assessment for over 300 hot springs in the country. Subsequently detailed geological, geophysical and tectonic studies on several thermal provinces, geochemical characteristics of the thermal discharges and reservoir temperature estimations were carried out by the National Geophysical Research Institute and the IITs, especially at Puga valley in Ladakh and Tatapani in Chhattisgarh.

It is estimated that there is a technical potential to produce about 10000 MW of geothermal power in India. From the point of view of electrification of rural and remote areas, there is even greater potential in terms of the impact of geothermal power where even small projects of 5 kW can significantly change the economic situation and living standards. The Himayalan region is one of the most promising geothermal areas. It contains about 100 thermal springs and falls in one of the most tectonically active zones. In the 1980s, the first pilot binary 5 kW power plant was successfully operated by the Geological Survey of India at Manikaran in Himachal Pradesh which proved the power producing capability there. The Puga Valley geo-thermal field in cold desert Ladakh of Jammu and Kashmir has been estimated to have a temperature of 240°C at a depth of 2000 meters.

As per the report of International Association for Ladakh Studies in 2005, the Puga valley in the Changthang area of Leh may soon offer the best solution to the ever-looming power crisis in Ladakh.

A bed of innumerable hot springs, this area has been identified as a belt with a very high potential for generation of geo-thermal energy. Panamik in Nubra valley and Demchok in Changthang are other geothermal belts in Ladakh, which lies on the zone of collision between the Indian and the Eurasian plate. Ladakh is so rich in renewable energy sources that it has the potential to become the laboratory and demonstration site for non-conventional energy projects in entire India. Some estimates suggest potential of energy generation as much as 40 MW of electricity in Puga valley (Harinarayana *et al.*, 2006).

3. Solar Energy Potential of Ladakh

Villages in Ladakh are isolated, houses are scattered, resources are inadequate and options for electrification are limited. Introduction of appropriate and renewable energy technologies have huge potential in improving the living conditions of people in Ladakh. Solar photovoltics (SPV) establishments at different villages can solve the problem of energy requirements in this cold desert region. The 250 Kv Diesel Generator at Tangtse in Durbuk block of Leh district supplies electricity to three villages for domestic lighting. The diesel generator besides polluting the otherwise pristine and highly fragile environment consumes an average of 48,200 litres annually. In collaboration with India Canada Environment Facility (ICEF), Ministry of New and Renewable Energy (MNRE), Ladakh Autonomous Hill Development Council (LAHDC) and people of Durbuk block, Ladakh Ecological Development Group (LEDeG) has installed a 4 X 25kWp solar photovoltaic power plant at Tangtse, which completely replaced the existing Diesel Generator set. The solar photovoltaic plant is now managed by the local people through a cooperative society known as Renewable Energy Development Cooperative Limited (REDCO). SPV home lighting systems have ameliorated the situation of fossil fuel dependency considerably. LEDeG has installed smaller SPV power plants of 1kW or more at Lingshed School and LEDeG's offices at Karzoo and Ribook Centre. Till date LEDeG has distributed over 1015 solar home-lighting systems and 1420 solar lanterns in Ladakh under the MNRE's programme.

3.1 Passive Solar Architecture for Room Heating in Extreme Cold

With the temperature dropping down to -40°C, room heating primarily in the evening and early morning is not a choice but a matter of survival. The conventional methods of room heating is the use of kitchen stove and stove fuelled with dung, wood and kerosene which not only pollute environment but also cause human health hazard. Space heating through use of solar energy may be the alternatives. Trombay wall technology is an extremely simple technology built on the principle of passive solar gain and minimization of heat loss through various design and insulation techniques. Trombay house can reduce reliance on heating fuels by about two-third, besides reducing indoor air pollution and health hazards. LEDeG retrofitted 135 houses with Trombe wall in Durbuk block of Leh district.

4. Contribution of DIHAR towards Non Conventional Energy Sector

4.1 Installation of Solar Power Plant of 216KW Capacity

At DIHAR, currently total electricity is generated by diesel generators. The electricity is provided to the living accommodation, laboratories, for running equipments, for street/security lights *etc.* and burning of the diesel is causing air pollution to the environment. Solar radiation is one of the most abundant natural resources in Ladakh, with annual solar radiation exceeding averages for other areas of India with high insolation (Table 1) (Purohit and Purohit, 2010). In Ladakh region out of 365 days more than 300 days are sunny days with sufficient sunlight.

Table 1 : Solar radiation estimates for Leh-Ladakh compared with other cities in India that receive high insolation.

Location	Altitude (m)	Annual solar radiation on horizontal surface (kWh/m^2/year)		
		Global	Diffusive	Beam
Jaipur	390	2067	691	1396
Ahmedabad	55	2110	746	1364
Leh	3514	2149	695	1454

Considering the fact, DIHAR has taken up a project for installation of a 216 KWA solar power plant with minimum power output of 100 KW for 4.5 hrs backup time. The plant has been installed in DIHAR campus and it is providing electricity for the street/security lights as well as to solar controlled green house and solar hatchery. It will soon cater electricity demands for the administrative block and the laboratories for running various research projects.

Fig. 1 : The Solar Power Plant at DIHAR, Leh

DIHAR has also constructed passive solar architecture for room heating during extreme winter months. DIHAR has successfully constructed Trombay wall shed for keeping poultry birds in its research unit and it is found very effective when ambient temperature goes below -25°C during severe winter.

5.2 Biogas Technology

Biogas, which is generated from organic digestion under anaerobic conditions by mixed population of microorganisms, is an alternative energy source that is now being utilized both in rural and industrial areas. Biogas technology offers an attractive route to utilize certain categories of biomass for meeting partial energy needs. Unlike other forms of renewable energy, biogas does not have any geographical limitations and required technology for producing energy and it is neither complex nor monopolistic (Balat and Balat, 2009). Spearheading the effort to tap alternate sources of energy in remote areas, the Defence Institute of High Altitude Research (DIHAR) has commissioned a biogas plant at Leh-Ladakh at an altitude of 3500 m above mean sea level (Fig. 2). It has been set up in collaboration with the Bhabha Atomic Research Centre (BARC), Mumbai for various R&D purposes. The capacity of the biogas plant is 0.5 t/day and it will generate about 35-50 m^3 of biogas per day during the processing of biodegradable waste. This can be used for cooking/boiler purposes [1.5-2.1 m^3 of biogas (depending on the methane content of 50-65%) equivalent to 1 liter of diesel in terms of heat output].

Fig. 2 : Biogas Plant at DIHAR, Leh

5.3 Research on Biodiesel

The major part of all energy consumed worldwide comes from fossil sources (petroleum, coal and natural gas). However, these sources are limited, and will be exhausted in the near future (Dermibas, 2005). The world is presently confronted with the twin crises of fossil fuel depletion and environmental degradation. Indiscriminate extraction and lavish consumption of fossil fuels have led to reduction in underground-based carbon resources. Since the oil crisis of the 1970s and recognition of the limitations of world oil resources, vegetable oils have received special attention (Pratt *et al.,* 2002). The search for alternative fuels, which promise a harmonious correlation with sustainable development, energy conservation, efficiency and environmental preservation, has become highly pronounced in the present context. Biodiesel is among those alternatives and it is defined as an alternative fuel for diesel engines produced by chemically reacting a vegetable oil or animal fat with an alcohol such as methanol or mono-alkyl esters of long chain fatty acids derived from a renewable lipid feed stock, such as vegetable oil or animal fat (Ma and Hanna, 1999; Van Gerpen, 2005; Wang *et al.,* 2006). Usage of biodiesel will allow a balance to be sought between agriculture, economic development and the environment (Meher *et al.,* 2006). In India, biodiesel is mainly produced from *Jatropha curcas* as a fuel substitute. Using this non-edible oil for making biodiesel has a great opportunity of making it economically feasible (Tiwari *et al.,* 2007). Generally a blend of 5 percent to 20 percent biodiesel is used in India (B_5 to B_{20}). The use of biodiesel from *Jatropha curcas* at the village level for operating oil engines used for pumping water, running small machinery and generating electricity is another possibility.

Ladakh remains cutoff from rest of the world during winter months for about six months and hence diesel fuels have to be stored for a long period of time to avoid shortage of fuel during winter. In this region mercury drops to -25 to -30°C during winter period. Due to very low temperature, the usage of petroleum diesel in the vehicles leads to many practical problems. Keeping this in view, (DIHAR) has started work on storage study of biodiesel and its blends, physico-chemical property analysis and performance evaluation under sub-zero cold desert conditions of Ladakh.

Storage studies on biodiesel and its blends (B_1 to B_{20}) with winter diesel (80% AT F + 20% DH PP) have been carried out in open environmental conditions in Ladakh in winter months under subzero temperature upto -20°C. B_{100} served as control treatment. Physico-chemical analysis of the Biodiesel blends (B_1 to B_{100}) with winter diesel was carried out for the following characteristics: cloud point, pour point, kinematic viscosity at -5°C and calorific value of different biodiesel blends. Glycerol was used in different molar concentrations in one liter of ATF and biodiesel blends to evaluate antifreezing characteristics of biodiesel blends in subzero temperature conditions.

During the storage study it was observed that treatments of biodiesel blends (B_1 to B_{20}) with winter diesel does not freeze upto -17°C but beyond that temperature, all the treatments got frozen under open environmental conditions. The viscosity data (Fig. 3) showed that a slight increase in viscosity was observed after the addition of biodiesel blends from B_{10} to B_{100} and were ranging from 2.85 to 4.4 cSt. The specification of Kinematic viscosity is in this range of 1.9-6.0 cSt in the ASTM D6751 biodiesel standard (Gerhard and Kevin, 2005). Results showed that the cloud point of winter diesel was -12°C, while blending of biodiesel with winter diesel (B_{10} to B_{100}) resulted in gradual increase in cloud point of the blend and neat biodiesel (B_{100}) having cloud point of +1°C (Fig. 4). Pour point of pure winter diesel was the lowest (-16°C) and blending with biodiesel (B_{10} to B_{100}) resulted in gradual increase in pour point and neat biodiesel showed highest pour point i.e. 0°C (Fig. 5). Antifreezing study showed that biodiesel blends with ATF did not freeze upto B_{30} with addition of glycerol, but without adding glycerol, biodiesel blends got frozen beyond B_{20}. Calorific value of different biodiesel blends was also determined and from the analysis slight decrease was found in calorific value of biodiesel blends after B_{30} (Fig. 6).

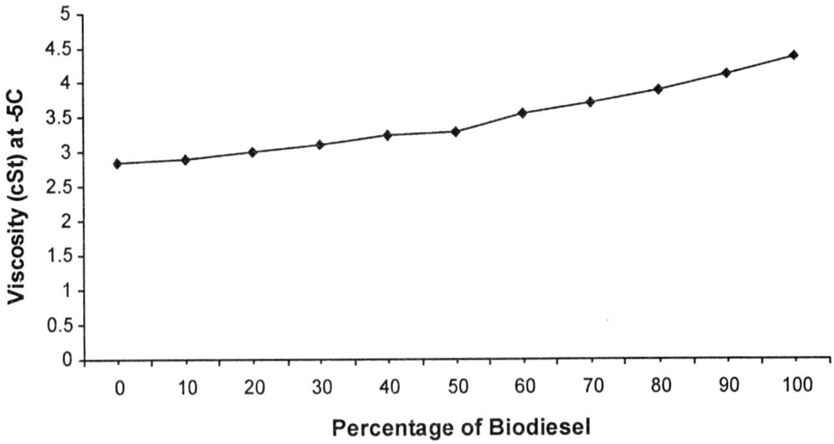

Fig. 3 : Change in viscosity at 5°C with increasing percentage of biodiesel in winter diesel

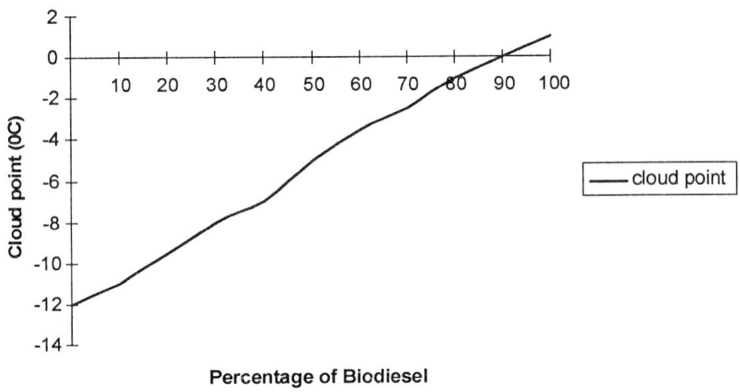

Fig. 4 : Change in cloud point with increasing percentage of biodiesel in winter diesel

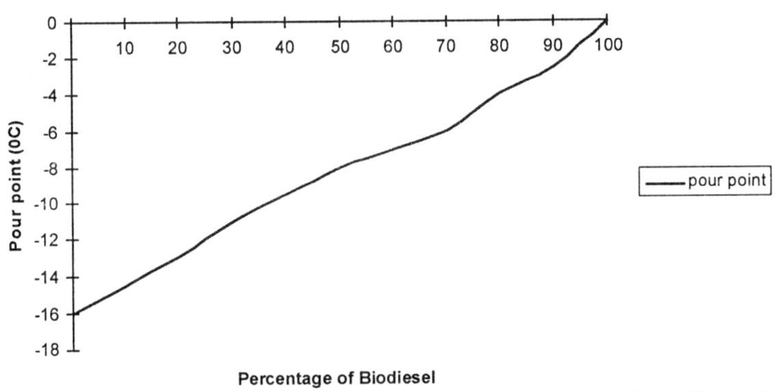

Fig. 5 : Change in pour point with increasing percentage of biodiesel in winter diesel

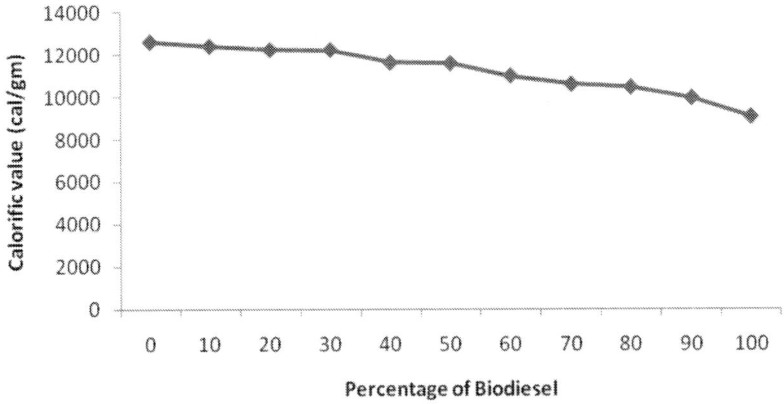

Fig. 6 : Calorific value of biodiesel blends

From the above study, it is concluded that biodiesel blends (upto B_{20}) with winter diesel can be stored upto -17^{0}C under open environmental condition of Ladakh. Cloud and pour point vary (increases) with the amount of biodiesel blended with petroleum diesel fuel (winter diesel) but it is within the safe limits as there is not much appreciable risk of fuel filter blockages. Viscosity increases as the concentration of biodiesel increases but the viscosity is closer to that of petrodiesel (winter diesel).

6. Conclusion

Installation of 216 kw solar power plant will prove to be a milestone towards the utilization of solar energy as the plant will cater most of the energy requirements of the institute. Biogas plant though commissioned for demonstration purposes, has paved the way for exploitation of the biogas energy in the entire Ladakh region. Lots of difficulties are experienced while using the petrodiesel particularly during winter months in the 'L' sector which can be avoided by blending the petrodiesel with biodiesel. By introducing antifreezing agents into biodiesel blends, freezing characteristics are greatly improved. Using biodiesel will also have various environmental benefits. In future projects may be undertaken to explore the utilization of geothermal energy in the Puga valley of Ladakh.

References

Balat, M. and Balat, H. (2009). Energy Sources, Part A: Recovery, Utilization, *Environmental Effects*, 31(4): 1280-1293.

Dermibas, A. (2005). Biodiesel production from vegetable oils via catalytic and noncatalytic supercritical methanol transesterification methods. Prog. *Profressive Energy Compusting,* 31: 466-487.

Gerhard, K. and Kevin, R.S. (2005). Kinematic viscosity of biodiesel fuel components and related compounds. Influence of compound structure and comparison to petrodiesel fuel components. *Fuel,* 84(9): 1059-1065.

Harinarayana, T., Abdul Azeez, K.K., Murthy, D.N., Veeraswamy, K., Eknath Rao, S.P., Manoj, C. and Naganjaneyulu, K. (2006). Exploration of geothermal structure in Puga geothermal field, Ladakh Himalayas, India by magnetotelluric studies. *Journal of Applied Geophysics,* 58: 280-295.

Ma, F. and Hanna, M.A. (1999). Biodiesel production: A review. *Bioresource Technology,* 70 (1): 1-15.

Meher, L.C., Vidya Sagar, D. and Naik, S.N. (2006). Technical aspects of biodiesel production by transesterification–a review. *Renewable and Sustainable Energy Reviews,* 10: 248-268.

Pratt, J.H., Henry, E.M.T., Mbeza, H.F., Mlaka, E. and Satali, L.B. (2002). Malawi Agroforestry Extension Project Marketing & Enterprise Program Main Report. Publication No. 47, Malawi Agroforestry.

Purohit, I. and Purohit, P. (2010). Techno-economic evaluation of concentrating solar power generation in India. *Energy Policy,* 38: 3015-3029.

Tiwari, A.K., Kumar, A. and Raheman, H. (2007). Biodiesel production from Jatropha oil (*Jatropha curcas*) with high free fatty acid: An optimized process. *Biomass and Bioenergy,* 31(8): 569 -575.

Van Gerpen, J. (2005). Biodiesel processing and production. *Fuel Proc. Tech.,* 86 (10): 1097-1107.

Wang, Y., Ou, S., Liu, P., Xue, F. and Tang, S. (2006). Comparison of two different processes to synthesize biodiesel by waste cooking oil. *Journal of Molecular Catalysis and Analytical Chemistry,* 252(1-2): 107-112.

□□□

Chapter-21

Challenges at High Altitude:
The Limits to Human Performance

Kalpana B.Hota, Sunil K.Hota, Priyanka Dhar, O.P. Chaurasia and R.B. Srivastava
Defence Institute of High Altitude Research (DIHAR), DRDO, Leh, Ladakh

1. Introduction

The misty and ice-clad peaks of the Himalayas and the scenic horizons of the mountains often unveils an instinct of beginning a journey to vanquish and conquer the heights of glory and reach the pinnacle of success. High altitude biology has seen a tremendous relevance in the recent years not only as a result of operations by the Indian and US army but also due to conquest of mountain summits by mountaineers and sojourners. High altitude is essentially defined as a region elevated from sea level with semi-attenuated atmosphere, low atmospheric temperature and barometric pressure, high atmospheric aridity and extremely harsh environmental conditions. Although most of the undesirable effects of high altitude results due to reduction in atmospheric pressure and partial pressure of oxygen, additional deterioration results from extreme cold, exposure to ionizing radiation and high wind velocity.

Ascent to high altitude leads to decreased partial pressure of oxygen in the atmosphere and resultant decrease in alveolar partial pressure of oxygen. This reduces oxygen diffusion to the pulmonary artery and causes subsequent reduction in oxygen saturation in the capillaries. Such a hypoxic condition resultant of ascent to high altitude is often referred to as 'Hypobaric Hypoxia' (Williams *et al.*, 1977). Though most of the environmental hazards at high altitude could be metered well with proper food, clothing or shelter, hypobaric hypoxia remains unavoidable and limits the highest acceptable level of inhabitation, which for sea-level visitors to altitude is 4600-4900m, while for high altitude natives it is 5800-6000m. Heights greater than 8000 m are well above the highest human inhabitation and are inhospitable for a sustained human life (West, 1986). The primary limiting factor is the barometric pressure, which declines exponentially with altitude (West, 1996). This in turn results in decreased partial pressure of oxygen in the atmosphere leading to tissue hypoxia.

2. Acclimatization and Physiological Responses to High Altitude

By definition, acclimatization describes a chronic adaptation response by the body to allow for improved tolerance to altitude changes. While some studies indicate a 12-14 day period up to an altitude of 2300 m, others suggest that this process of acclimatization may take several months (McArdle *et al.*, 2001; Robergs and Keteyian, 2003). Within the process of acclimatization several systems respond distinctly for improved and enhanced work ability.

2.1. Cardiovascular Adaptations to Altitude

With acclimatization there is a reduction in resting and submaximal heart rate indicating a return to normal homeostasis within this system. Mean arterial pressure increases due to systemic vascular resistance, increased catecholamine secretion and increased blood viscosity resulting from increased hematocrit. The reduced myocardial response during vigorous work at altitude is a resultant of decreased plasma volume, increased total peripheral resistance and an increase in parasympathetic tone decreasing maximal heart rate. While it seems that maximal cardiac output decreases along with muscle blood flow with chronic altitude exposure at 4300 m, VO_{2max} values increase after acclimatization.

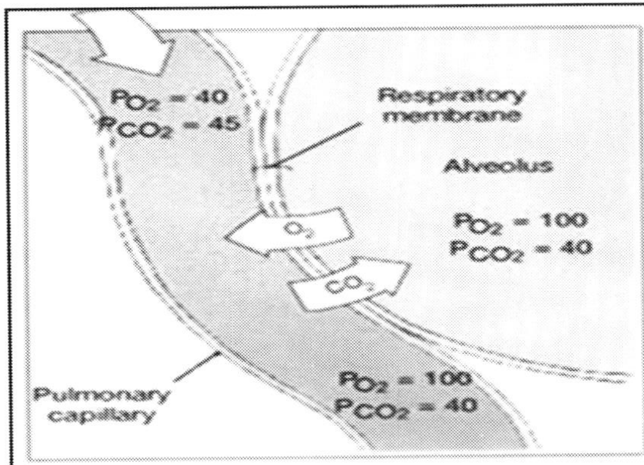

Fig. 1 : Partial pressure and diffusion in lungs

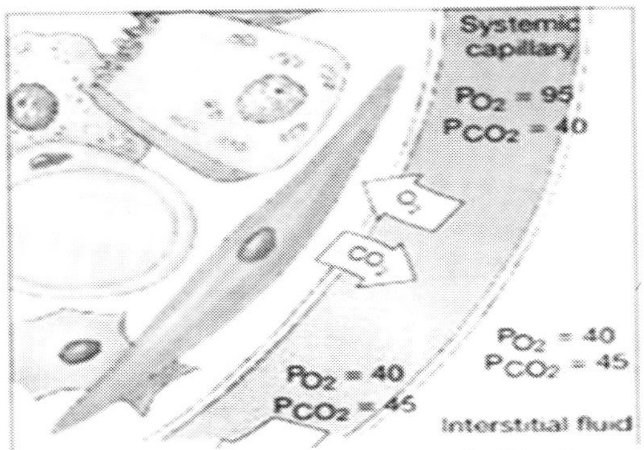

Fig. 2 : Partial pressure and diffusion in peripheral tissue

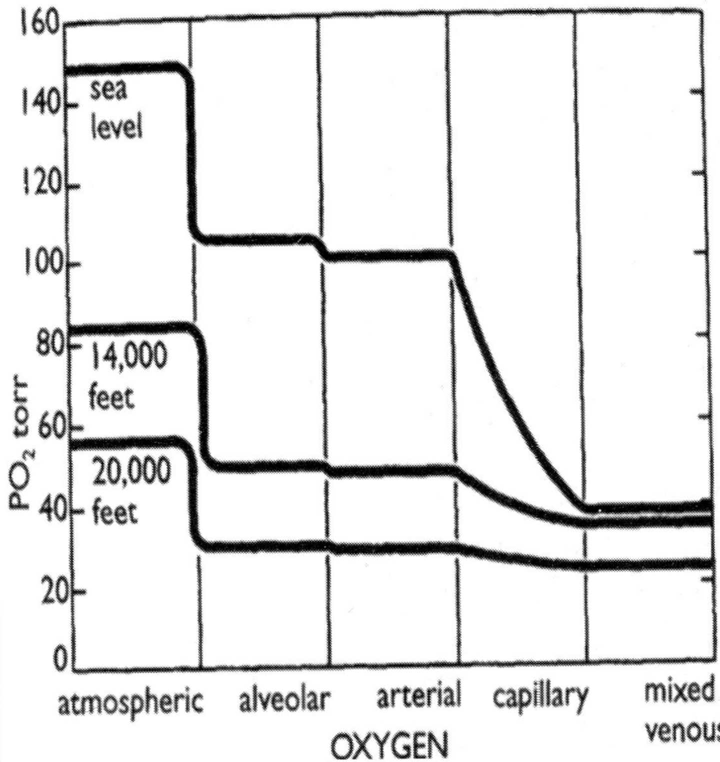

Fig. 3 : Oxygen Cascade at different altitudes (Courtsey : Heath and Williams ; 1977)

2.2. Pulmonary and VO_2 with Chronic Exposure to Altitude

Chronic exposure to high altitude results in increased pulmonary blood pressure and vascularity leading to improved pulmonary perfusion (Brooks, 2000). Ventilation continues to be elevated with acclimatization and may be an indication of increased chemoreceptor sensitivity to blood gas changes occurring at altitude. In addition, hyperventilation allows for normalization of alveolar and arterial oxygen pressures and reduced PCO_2 levels for acid-base balance.

2.3. Hematological and Muscular Adaptations to Chronic Exposure to High Altitude

Exposure to high altitude leads to several hematological and muscular adaptations that continue during the acclimatization process allowing for increased tolerance to hypoxic conditions. As mentioned above, there is an increase in hemoglobin, red blood cells and hematocrit. In addition, with the increased red blood cells 2,3-DPG, the oxygen dissociation curve shifts to the right indicating a facilitated release of oxygen to the working tissues. Facilitating the oxyhemoglobin dissociation shift is an increase in chemoreceptor control of ventilation brought on by decreased bicarbonate in the cerebrospinal fluid and excretion of bicarbonate by the kidneys (Brooks, 2000). With acclimatization there is also an increase in skeletal muscle vascularity and tissue myoglobin to provide improved oxygen transport and cellular oxygen transport, respectively.

2.4. Bioenergetics and Performance Adaptations to Chronic Exposure to High Altitude

After prolonged exposure to high altitude, muscle glycogen stores are reduced with an concomitant increase in blood glucose utilization (Robergs and Keteyian, 2003). It has been reported that with prolonged exposure there is a drop in circulating blood lactate levels during periods of exertion. Recent studies indicate that after acclimatization there is an increase in lactate uptake by active and inactive skeletal muscles, heart, kidney and liver (Brooks, 2000). Additionally, the reduced ability of the central nervous system and cardiovascular system in hypoxic environments reduces the level of work the body can reach for extended periods of time.

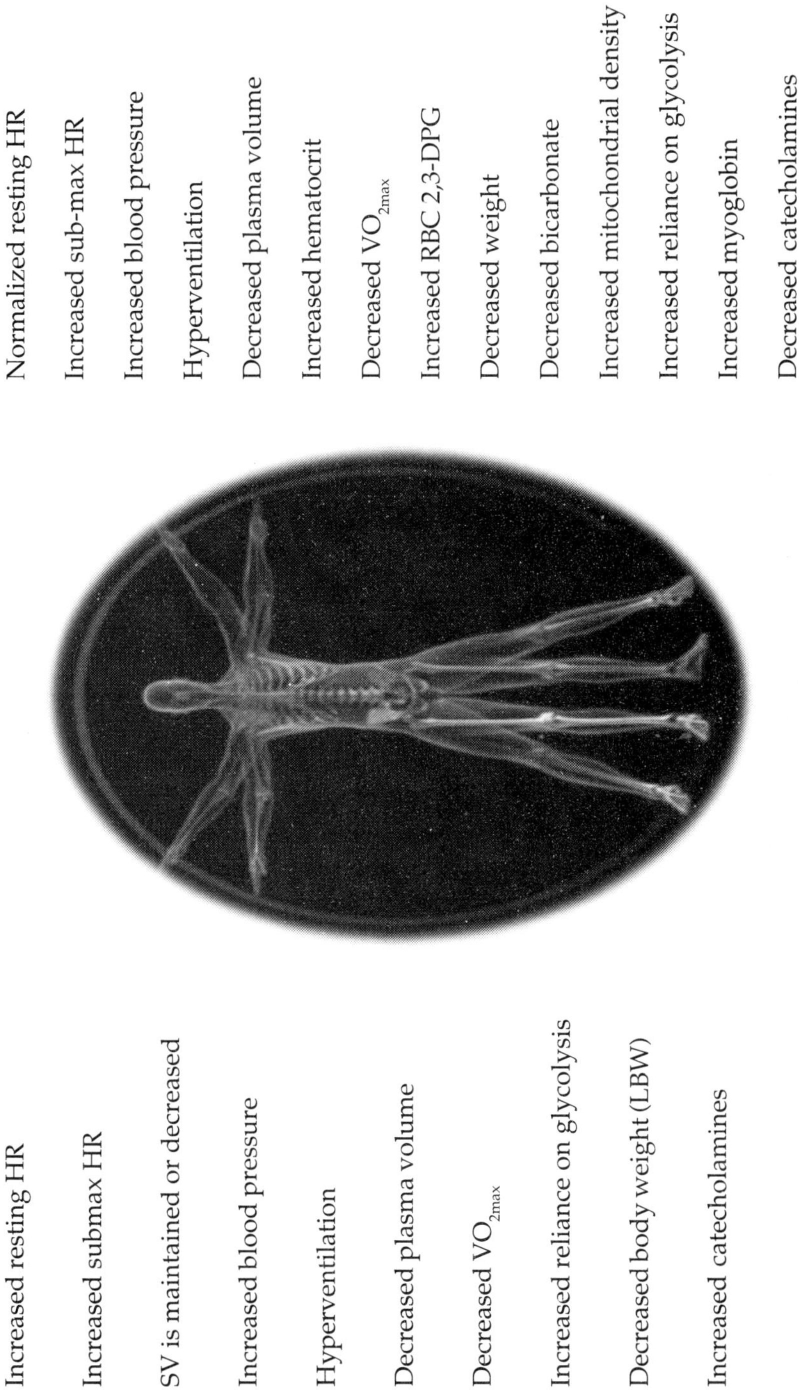

Fig. 4 : Physiological Reponses to Acute and Chronic Exposure to Altitude.

It has also been reported that with prolonged exposure to high altitude there is weight loss and associated changes in body composition. Both losses in lean tissue and body fat have been reported to be in direct association to increased elevation (McArdle *et al.,* 2001). This may be due to reduced appetite experienced at increasing altitudes. Fat catabolism may also increase along with gluconeogenesis if diet is inadequate. Furthermore, for any given workload, the level of exertion is also increased as the partial pressure of oxygen is reduced. This is reflected in the decline in maximal oxygen consumption (VO_{2max}) seen with increasing altitudes.

In general, even after acclimatization to altitude the greater the distance to be covered in performance the greater the time to achieve that distance as altitude is increased. There is also a reduced plasma volume when combined with increases in hematocrit that would increase blood viscosity and reduce oxygen transport capabilities. Reported decreases in bicarbonate (HCO^{3-}) would result in decreased lactate efflux to the blood and subsequent decrease in muscle tissue pH. This of course would lead to an earlier onset of fatigue (Brooks, 2000). Lastly, the hyperventilatory response at altitude adversely affects performance causing increased work of breathing there by leading to an earlier onset of fatigue.

3. Problems at High Altitude

3.1. Acute Mountain Sickness (AMS)

AMS refers to a self limiting condition affecting previously healthy individuals going rapidly to high altitude. The most common symptoms are headache, nausea, anorexia, insomnia, fatigue/lassitude, vomiting and dizziness (Singh *et al.,* 1969; Hackett *et al.,* 1976; Hackett and Roach, 2000). The generally accepted Lake Louise scoring system requires the presence of headache and at least one of the other symptoms, rated in severity on a scale of 1 to 3 (Roach *et al.,* 1993). These symptoms may worsen over a day or two and may be worst on arising in the morning, but if the subject does not ascend further, the symptoms usually subside on their own. The illness may recur, however, if the so-journer ascends to a higher altitude.

Many physiological events associated with the pathophysiology of established AMS documents hypoventilation (Moore *et al.,* 1986), impaired gas exchange (interstitial pulmonary edema) (Larson *et al.,* 1982; Ge *et al.,* 1997), fluid retention and redistribution (Bärtsch *et al.,* 1992; Hackett and Roach, 2000) and increased sympathetic drive (Bärtsch *et al.,* 1991).

3.2. High Altitude Pulmonary Edema (HAPE)

HAPE typically occurs at an altitude of 3,000 m and is a major contributor of most deaths from altitude illness. As with AMS, it affects previously healthy individuals ascending to high altitude. It may be preceded by AMS, and its symptoms typically begin 2 to 4 days after arrival at altitude. HAPE begins with a subtle nonproductive cough and shortness of breath, both at rest and especially with attempts to modest exercise, and progresses to dyspnea, even at rest, and a cough productive of pink, frothy sputum. Subjects are tachypneic and tachycardic, need to rest frequently, and have crackles on chest auscultation. If the condition worsens, lethargy, coma, and death may ensue. Men seem to be more susceptible to HAPE than women.

3.3. High Altitude Cerebral Edema (HACE)

HACE is a rare, life threatening form of altitude illness. It shares almost similar pathophysiology as that of AMS, and it is suggestive that most people with HACE have had preceding symptoms of AMS. HACE occurs at an altitude above 2500 m, but is much more common above 4000 m. An epidemiological study among Indian military found that as many as 1-2% of those ascending to an altitude of 4500 m developed HACE (Singh *et al.*, 1969). The symptoms include, in order of appearance: Confusion, changes in behavior, fatigue, ataxia, difficulty in speaking, vomiting, hallucinations, seizure and coma.

The underlying mechanism of HACE remains unknown. However, increased cerebral blood flow (CBF) and capillary leakage of fluid occur, resulting in an increase in cerebral blood volume and parenchymal volume (Hackett, 1999). The blood-brain barrier and intracranial hemodynamics are the two key elements in the pathophysiology of HACE and AMS. The mechanism of capillary leakiness remains unknown but may be linked with high capillary pressure causing vessel wall stress, or various substances, released in response to hypoxia, that are active on the endothelium, including histamine, arachidonic acid, bradykinin, nitric oxide and vascular endothelial growth factor (VEGF) (Severinghaus, 1995). Treatment is directed both to lowering intracranial pressure by reducing the volume of intracranial contents, and to stopping the vasogenic leak.

3.4. Frost Bite

Frostbite, medically named cryopathy, sets in when the body is exposed to temperatures of, or below, $0^{0}C/32^{0}F$. The effects can be seen most commonly in the fingers, toes, chin, cheeks, nose and earlobes, but it is actually the skin tissue and blood vessels of these extremities which are damaged. The air at high altitude is dry and damages the mucous membrane which is kept moist by the air we breathe resulting in loss of body fluids. There is more loss of fluids in the body as one tends to sweat a lot while climbing and level of fluid intake also reduces. All these factors result in blood viscosity. Due to viscosity not enough oxygen is supplied to the capillaries and the body gets cold resulting in the frostbite.

Frostbite is said to occur in three stages. The speed at which they occur depends on the climatic conditions and the duration for which one is exposed. First, the skin turns pale and is very soft to touch and the sensation is like that of a pin prick. These symptoms are easily reversible with proper care. The second stage is that of superficial frostbite and blistering occurs. The affected area becomes yellowish grey in colour and the skin feels leathery. If these are not treated properly, deep frostbites may occur which can cause permanent damage. This happens because at extreme temperatures the blood vessels can get frozen.

The pathophysiology behind frostbite includes generation of oxygen free radicals, production of prostaglandins and thromboxane A2, release of proteolytic enzymes, and generalized inflammation. Tissue injury is greatest when cooling is slow, cold exposure is prolonged, rate of rewarming is slow, and, especially, when tissue is partially thawed and refreezes (Friedman and Kritzler, 1947).

In spite of the precaution, if one gets frostbite immediate medical assistance should be taken. Till medical attention is not available the patient should be brought down to lower altitudes and the affected areas should be dipped in warm water for about twenty minutes. Dry heat or very hot water should not be used. The skin should be kept clean and above 5000 m oxygen can be given. Medical advice and treatment should be taken if the frostbites is not treated in the initial stages. Complications can occur and they can become extreme and can also cause loss of limbs and permanent damage.

3.5. Dehydration

At high altitude low air pressure results in more rapid evaporation of moisture from skin surface and lungs. High altitude areas are also very low in humidity, which means evaporation is further accelerated.

The combination of these two factors results in dehydration. The signs of dehydration include: lack of perspiration, dizziness, shortness of breath, nausea, headache, fatigue. Symptoms of headache, fatigue, or dizziness may appear first, and may be mistaken for altitude sickness (Cherniakov and Maksimov, 1967). Drinking sufficient water is the only means to avoid dehydration.

3.6. Sleep Disturbances

Sleep disturbances are very common at high altitude. There is a decrease in both deep and rapid eye movement (REM) sleep, with frequent arousals. Some of these arousals are due to periodic breathing (Cheyne-Stokes respiration), which is common at high altitude, and does not appear to be related to AMS. Some people are unaware that they have periodic breathing while others wake in a panicked state following an apnoeic episode, thinking that they are being suffocated. Although not investigated extensively, especially following ascent to high altitude where central sleep apnea commonly develops, breathing pattern and brain blood flow are closely linked by partial pressure of arterial CO_2 ($PaCO_2$) (Topor *et al.*, 2004, Xie *et al.*, 2005). The effects of CO_2 on CBF provide an important counter regulatory mechanism, which serves to minimize changes in brain thereby stabilizing the breathing pattern in face of perturbations in $PaCO_2$ (Xie *et al.*, 2005). Hypocapnia normally causes marked cerebral vasoconstriction and reduces CBF, which attenuates the fall in brain PCO_2 relative to that of the arterial blood (Dempsey, 20005). Accordingly, ventilatory inhibition in response to reduced $PaCO_2$ will be lessened because of the attenuated decrease in central chemoreceptor. Changes in CBF may, therefore, play an important role in etiology of central sleep apnea at high altitude.

3.7. Cognitive Impairment

Much early work by McFarland documented the effect of hypoxia on mental performance at high altitude (MacFarland, 1937). On the basis of observations and tests of sensory, motor, and cognitive function, McFarland observed that individuals taken rapidly (hours) to 4,000-4500 m exhibited impairment in both simple and complex psychological performance. Motor functions, such as handwriting, were also impaired but to a lesser extent, and sensory modalities were affected little if at all several. Investigators have documented decrements in performance on a variety of neuropsychometric tests for cognitive and motor functions after sudden exposure to even relatively moderate hypoxia (2000–4500 m) (Stickney and Van Liere, 1953; Ernsting, 1978).

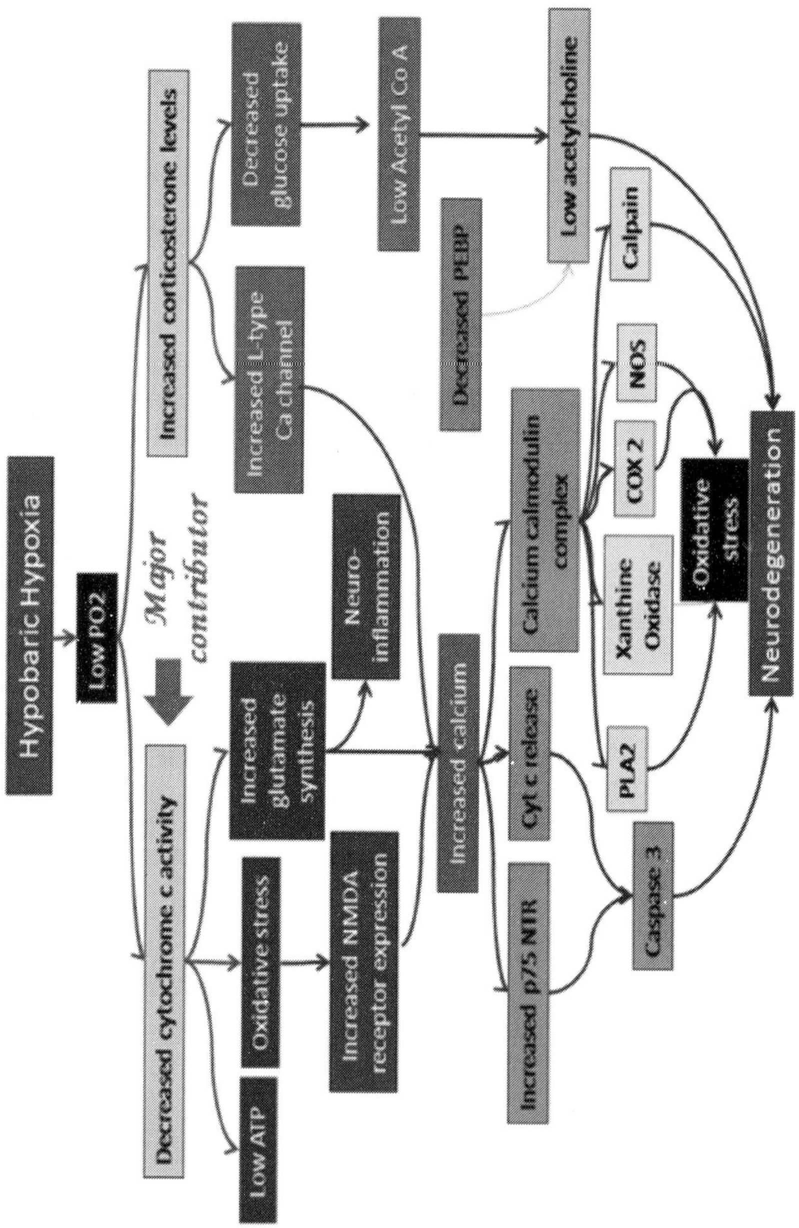

Fig. 5 : Molecular mechanism pertaining Hypobaric Hypoxia induced memory impairement.

Changes in a visual-positioning test performed during light work have also been reported at an altitude as low as 1500 m (Denison *et al.*, 1966). Acute exposure to hypoxia (few minutes) at altitudes above 6500 m is known to cause severe neurobehavioral dysfunctions and loss of consciousness in non-acclimatized individuals (Hackett, 1999). These changes with acute hypoxia are evidence that even modest levels of hypoxia can impair brain function.

A recent study in humans has shown that 15 adults (29–37 years old), tested under high altitude conditions (4500 and 5050 m), displayed difficulties in recalling word lists, specifically those words that came early in the list (primacy effect). In this study, memory recall remained impaired 45 days after descent from high altitude (Pelamatti *et al.*, 2003). There is further evidence that memory impairment may last several months after returning to the lowlands. A number of other studies have shown that verbal and visual short-term memory capacity and recall is impaired at altitudes starting at 2500 m (Hopkins *et al.*, 1995; Phillips and Pace, 1966). Recent studies by our laboratory on large populations of acclimatized low landers have revealed a duration dependent impairment in cognitive functions along with changes in alpha wave rhythms in the temporal region. Altered autonomic responses have also been observed in low lander subjects staying at high altitude for 6 months. These studies suggest that a critical impairment in higher cognitive functions is one of the earliest and most insidious consequences of exposure to high altitude.

The mechanisms pertaining to hypobaric hypoxia induced cognitive impairment appear to be multifactorial and may involve oxidative stress, neurotransmitter alterations, altered bioenergetics, altered neuromorphology and disturbed ionic homeostasis. Animal studies with rats exposed to hypobaric hypoxia reveal occurrence of oxidative stress, neuronal degeneration and dendritic atropy (Titus *et al.*, 2007, Hota *et al.*, 2007, 2008 a, b, c, Barhwal *et al.*, 2007, 2008, 2009 a,b). Though the mechanisms pertaining to free radical generation and neuromorphology in ischemic conditions have been considerably studied, the mechanisms involved in hypobaric hypoxia remain to be investigated. Previous studies have also indicated towards the occurrence of glutamate excitotoxicity in hypoxic and ischemic stress (Goldberg *et al.*, 1987; Hemi *et al.*, 2003, Hota *et al.*, 2008). Animals exposed to simulated hypobaric hypoxia showed robust calcium influx into neuronal cells through the NMDA receptors and L type calcium channels, thus mediating excitotoxic cell death (Gwag *et al.*, 1999, Sunil *et al,* 2010, Barhwal *et al.,* 2009 a). Recent findings also reveal both mitochondria dependent and mitochondria independent mechanisms of neurodegeneration following hypobaric hypoxic exposure which is accompanied by changes in acetylcholine and glutamate mediated signaling and alterations in transcriptional regulation of receptor proteins (Hota *et al.,* 2008, 2009, Kalpana *et al.,* 2010, Muthuraju *et al.,* 2009).

3.8. Oxidative Stress at High Altitude

Several studies have indicated the association of oxidative stress with stress related physiological disorders. 'Oxidative stress' refers to the situation of a serious imbalance between production of free radicals and the antioxidant mechanisms involved in quenching of the free radicals (Halliwell, 2007). A disturbance in the pro-oxidant-antioxidant balance in favor of the former, leads to potential oxidative damage. Oxidative stress may be an outcome of diminished antioxidants or increased production of reactive oxygen species. Major sources of free radicals in stress is the leakage of electrons from the electron transport chain in mitochondria, increased activity of pro-oxidant enzymes, increased stress hormones, calcium mediated mechanisms and lysosomal sources. However, the existence of an efficient antioxidant defense mechanism in living systems protects the cellular components from free radical damage in conditions of oxidative stress. Antioxidants are molecules that slow or prevent the oxidation of other molecules. They terminate free radical mediated chain reactions by removing radical intermediates and getting self oxidized (Halliwell, 2001; Halliwell, 1999). Glutathione (g- glutamylcysteinglycine), a tripeptide consisting of glutamic acid, cysteine and glycine is one of the principal antioxidants in living cells. Quenching of free radicals by glutathione is mediated by the antioxidant enzyme Glutathione peroxidase, Catalase Glutathione reductase and Superoxide Dismutase (Gutteridge and Halliwell, 1989).

The free radicals generated as byproducts of cellular metabolism are therefore adequately quenched by the antioxidant system. However, excessive generation of free radicals or impairment of the antioxidant mechanisms during stress disturbs the delicate balance that protects cells from oxidative damage thus culminating in cell death. Increased generation of free radicals along with decreased antioxidant levels, triggers oxidative damage of biological macromolecules like lipids, proteins and nucleotides in stressful conditions. Oxidative stress mediated damage to lipids, proteins and DNA has been reported in high altitude exposure (Magalhaes *et al.*, 2004; Moller *et al.*, 2001). Membrane lipids in particular are highly susceptible to oxidative stress. Peroxyl radicals attack PUFA side chains generating new C· radicals to propagate a chain reaction and a lipid hydroperoxide (LOOH). The overall effects of lipid peroxidation are increased membrane fluidity making easier for phospholipids to exchange between the two halves of the bilayer, increase the leakiness of the membrane to substances that do not normally cross it other than through specific channels and damage membrane proteins inactivating receptors, enzymes and ion channels (Halliwell and Chirico, 1993). Free radicals can also potentially damage cellular DNA resulting in replicational and transcriptional changes in living cells. ·OH can add on to guanine and adenine forming adducts that can have various fates. Pyrimidines are also attacked by OH to give multiple products. In addition to nitrogenous bases all positions in the sugar molecule are susceptible to the formation of carbon centered radicals due to hydrogen abstraction by ·OH. Damage to proteins can occur by direct attack of ROS/RNS upon them or by secondary damage involving attack by end products of lipid peroxidation like MDA.

Human studies on low landers inducted to high altitude invariably showed increase in free radicals and depletion of antioxidants in the serum. The brain is especially vulnerable to free-radical damage for a number of reasons. First, it has one of the highest concentrations of polyunsaturated fat contents that are especially prone to lipid peroxidation. Second, the brain has a very high oxygen demand. It accounts for 20 percent of the body's oxygen requirement, even though it makes up only 2-4 percent of the body weight. Third, the nervous system, especially the brain and spinal cord, undergoes a progressive accumulation of tissue iron levels which is a pro-oxidant and is poor in endogenous antioxidants as well (Dawson and Dawson, 1996).

4. Therapeutics and Prophylactics for High Altitude

With the research on high altitude physiology gaining momentum in recent years and as a consequence of a concerted effort of physiologists from all over the world, several therapeutics and prophylactics have been recommended for high altitude related maladies. The most common therapeutics used for the management of altitude illness in normal individuals are presented in Table 2 and have been reviewed elsewhere (Hackett and Roach, 2001).

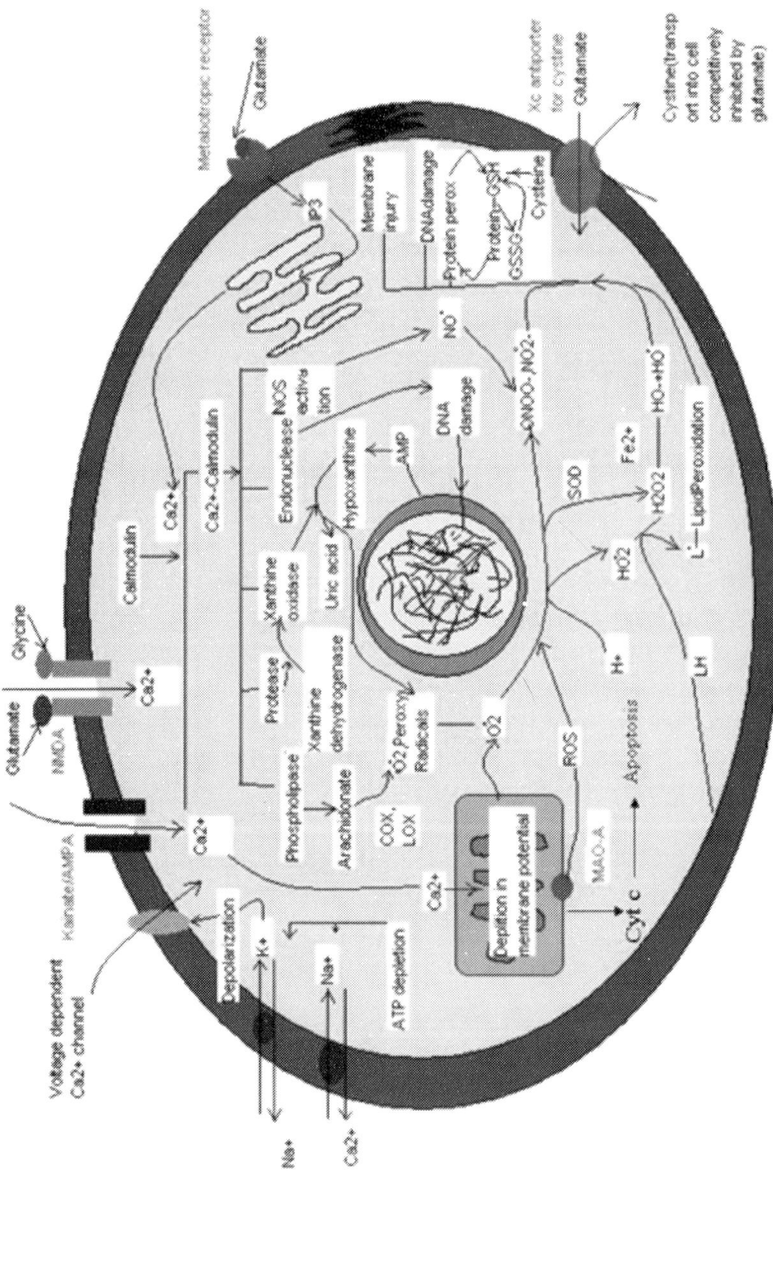

Fig. 6 : Schematic diagram showing Biochemical mechanism for generation of free radicals in Hypoxia.

DO's	DONT's
Do take plenty of water (300 ml/hr)	Don't take anything which will slow down breathing such as alcohol, sleeping pills, narcotics, pain medication, tobacco.
Do wear multiple layers of warm clothing	Don't wear wet clothing
Do eat high carbohydrate low fat meals	Don't over-exert
Do moderate physical activity, take rest when you feel tired	Don't Smoke
Do use sunscreen (SPF > 35) and wear UV protective sun glasses	Don't touch metals with bare hands
Do wear properly fitted shoes and gloves	Don't neglect health complications
Do wash feet with warm water and change socks on washing feet	Don't cover face completely while sleeping
Do ensure ventilation when Bukhari is on	Don't take medicine without medical advice
Do take repeated light meals	Don't explore uninhabited areas alone
Do take cough candy and cover nose and mouth with scarf for air humidification	Don't climb more than 5000 ft/day
	Don't sit in cuddled postures for long

Table 1 : DOS and Dont's at High Altitudes

Table 2 : Medical therapies and therapeutics employed in the treatment and management of high altitude illnesses

AGENT	INDICATION	DOSE	MECHANISM OF ACTION	ADVERSE EFFECTS	COMMENTS
Oxygen	All high-altitude (HA) illnesses	2-4 L/min by cannula then 1-2 L/min	Increase PaO_2 Reduces blood flow, CBV, pulmonary artery pressure	None	Lifesaving for HAPE; improves headache within min in AMS
Portable Hyperbaric chamber	All high altitude (HA) illnesses	Depends on model, 2-4 psi for a minimum of 2 hr, cont. as long as necessary	Stimulates descent; increases PaO2	Potential rebound effect after removal of patient from chamber; limits access to airways	Effects equivalent to the administration of low-flow oxygen; can be lifesaving
Acetazola-mide	Prevention of AMS Treatment of AMS	125-250 mg orally twice a day before ascent & 1st 2 days at HA 250 mg twice a day orally until symptoms resolve	Carbonic anhydrase inhibitor; causes bicarbonate diuresis & respiratory stimulation, increases PaO2; reduces formation of CSF, promotes ion transport across BBB	Parasthesis ; alters taste of carbonated beverages; Polyuria	Sulfonamide reactions possible; can be taken episodically for symptoms; no rebound effect, avoided by breast feeding women

AGENT	INDICATION	DOSE	MECHANISM OF ACTION	ADVERSE EFFECTS	COMMENTS
Dexamethasone	Prevention of AMS	2 mg every 6 hr or 4 mg every 12 hr orally	Unknown; may reduce brain-blood volume; may prevent Blood-brain barrier leak (blocks VEGF, inducible nitric oxide, lipid peroxidation)	Mood changes; hyperglycemia; dyspepsia; rebound effect on withdrawl	Can be lifesaving for AMS or HACE; effects evident in 2-8hr; no effect on acclimatization; no value in HAPE, preferably avoided by women who are pregnant
	Treatment of AMS HACE	4 mg every 6 hr orally, IM,IV 8 mg initially, then 4 mg every 6 hr orally, IM or IV			
Furosemide	AMS Or HACE	80 mg orally every 12 hr for a total of 2 doses	Diuresis; decreases extracellular fluid; causes venodialatation	Hypovolemia; hypotension; hypokalemia	Currently out of favor, not recommended for prevention; not established for use in HAPE
Nifedipine	Prevention of HAPE	20-30 mg of extended release formulation orally every 12 hr	Calcium channel blocker; reduces pulmonary artery pressure	Reflex tachycardia; hypotension (uncommon)	No value in AMS or HACE; pregnancy category C; not necessary if supplemental oxygen available
	Treatment of HAPE	10 mg orally initially, then 20-30 mg of extended release, orally			

AGENT	INDICATION	DOSE	MECHANISM OF ACTION	ADVERSE EFFECTS	COMMENTS
NSAIDS			Inhibit Prostaglandins	Dyspepsia; gastrointestinal bleeding	No clinical trials of aspirin for treatment or ibuprofen for prevention; naproxen ineffective
Aspirin	Prevention of headache	325 mg orally every 4 hr for a total of 3 doses			
Ibuprofen	Treatment of headache	400 or 600 mg orally once; maybe repeated			
Ginkgo biloba	Prevention of AMS	80-120 mg orally twice daily	Unknown; may block inducible nitric oxide; an antioxidant oxygen radical scavenger; may block platelet activating factor	Occasional headache; rare episodes of bleeding	Requires further study; preparations vary; should not be used with antithrombotic agents; maybe used by women who are pregnant or breast feeding
Antiemitics	Nausea; vomiting		Phenothiazine; centrally acting	Extrapyramidal reactions; causes sedation	Pregnancy category C; use diphenhydramine intramuscularly for extrapyramidal reactions

Besides the above mentioned therapeutics, an aloe vera based formulation (Alocal) developed by Defence Institute of Physiology and Allied Sciences (DIPAS) at Delhi, has also emerged as an effective prophylactic and therapeutic for frost bite related problems. Several prophylactics and antioxidant supplements that include Herbal Tea, omega fatty acid rich soft gel capsules, seabuckthorn berry based health drinks and seabuckthorn, apricot and Rhodiola based herbal appetizers have been developed by Defence Institute of High Altitude Research and found to be quiet effective. Portable HAPO bags and Nitric Oxide breathing system developed by Defence Institute of Physiology and Allied Sciences and Defence Bio-engineering and Electro Medical Laboratory (DEBEL) are being used for medical resurrection from HAPO and HACO.

Apart from these medications, high altitude ailments can be avoided by simply following some rules as tabulated in Table 1.

5. Conclusion

The understanding on physiological and patho-physiological manifestations on high altitude exposure has grown considerably in the past few decades. Research on high altitude physiology is no longer confined only to the mountaineers or the Everest expeditions. With the increase in number of sojourners, high altitude related illness has emerged as a major health concern. Investigations on polymorphisms and genetic basis of adaptability in high altitude native populations, has gained momentum in recent years. The search for molecular markers to facilitate prognosis and determination of susceptibility to high altitude maladies and the quest for prophylactics to improve acclimatization has been a thrust area in high altitude research in recent years. However, the discovery of therapeutics for treatment of high altitude maladies like AMS and HACE needs to be expedited. Sleep apnea and hypophagia still poise a challenge for prolonged stay at high altitude. Besides that the delayed effect following de-induction from high altitude on human physiology and its role in accelerating ageing and neurodegenerative disorders still remains to be explored. The reports on effect of high altitude exposure on reproductive system and fetal rejection still remain controversial. Emerging challenges like deep vein thrombosis, upper respiratory tract infections and gastric motility also demand scientific research. With the growing interest on research on high altitude physiology world wide, the discovery of an elixir for high altitude maladies does not appear to be a distant reality.

References

Barhwal, K., Singh, S.B., Hota, S.K., Jayalakshmi, K. and Ilavazhagan, G (2007). Acetyl-L-Carnitine ameliorates hypobaric hypoxic impairment and spatial memory deficits in Rats. *European Journal of Pharmacology*, 570: 97-107.

Barhwal, K., Hota, S.K., Prasad, D., Singh, S.B. and Ilavazhagan, G. (2008). Hypoxia induced deactivation of NGF mediated ERK ½ signaling in hippocampal cell culture: neuroprotection with Acetyl-L-Carnitine. *Journal of Neuroscience Research*, 86: 2705-21.

Barhwal, K., Hota, S.K., Baitharu, I., Prasad, D., Singh and S.B., Ilavazhagan, G. (2009a). Isradipine antagonizes hypobaric hypoxia induced CA1 damage and memory impairment: complementary roles of L-Type Calcium channel and NMDA receptors. *Neurobiology of Disease*, 34: 230-44.

Barhwal, K., Hota, S.K., Jain, V., Prasad, D., Singh, S.B. and Ilavazhagan, G. (2009b). Acetyl-l-carnitine prevents hypobaric hypoxia-induced spatial memory impairment through extracellular related kinase-mediated nuclear factor erythroid 2-related factor 2 phosphorylation. *Neuroscience*, 161: 501-514.

Kalpana, Barhwal, Hota., Sunil, Kumar, Hota., Om, Prakash, Chaurasia and Shashi, Bala, Singh (2010). Acetyl-L-Carnitine-Mediated Neuroprotection During Hypoxia Is Attributed to ERK1/2-Nrf2-Regulated Mitochondrial Biosynthesis. *Hippocampus* (in press).

Bärtsch, P., Maggiorini, M., Schobersberger, W., Shaw, S., Rascher, W., Girard, J., Weidmann, P. and Oelz, O. (1991). Enhanced exercise-induced rise of aldosterone and vasopressin preceding mountain sickness. *Journal of Applied Physiology* 71: 136–143.

Bärtsch, P., Shaw, S., Wiedmann, P., Franciolli, M., Maggiorini, M. and Oelz, O. (1992). Aldosterone, antidiuretic hormone and atrial natriuretic peptide in acute mountain sickness. In *Hypoxia and Mountain Medicine* (eds. J. R. Sutton, G. Coates and C. S. Houston), VT: Queen City Press, Burlington, pp. 73–81.

Brooks, G.A., T.D. Fahey, T.P. White, and K.M. and Baldwin (2000). Exercise Physiology: Human Bioenergetics and Its Applications, Third Edition. Mayfield Publishing Company. Mountain View, CA.

Cherniakov I.N. and Maksimov I.V. (1967). Dehydration of the human organism at high altitudes. *Voen Med Zh.*, 3: 62–64.

Dawson, V.L. and Dawson, T.M. (1996). Free radicals and neuronal cell death. *Cell Death and Differentiation*, 3: 71–78.

Dempsey J.A. (2005). Crossing the apnoeic threshold: causes and consequences. *Experimental Physiology* 90: 13–24.

Denison, D.M., Ledwith, F. and Poulton, E.C. (1966). Complex reaction times at simulated cabin altitudes of 5,000 feet and 8,000 feet. *Aerospace Medicine*, 37: 1010–1013.

Ernsting, J. (1978). Prevention of hypoxia—acceptable compromises. *Aviation Space and Environmental Medicine*, 49: 495-502.

Friedman N.B and Kritzler R.A. (1947). The Pathology of High-Altitude Frostbite. *American Journal of Pathology* 23:173–187.

Ge, R. L., Matsuzawa, Y., Takeoka, M., Kubo, K., Sekiguchi, M. and Kobayashi, T. (1997). Low pulmonary diffusing capacity in subjects with acute mountain sickness. *Chest*, 111: 58–64.

Goldberg, M.P., Weiss, J.H., Pham, P.C. and Choi, D.W. (1987). N-methyl-D-aspartate receptors mediate hypoxic neuronal injury in cortical culture. *Journal of Pharmacological and Experimental Therapeutics*, 243: 784–791.

Gutteridge, J.M. and Halliwell, B. (1989). Iron toxicity and oxygen radicals. *Baillieres in Clinical Haematology*, 2: 195–256.

Gwag, B.J., Canzoniero, L.M., Sensi, S.L., DeMaro, J.A., Koh, J.Y., Goldberg, M.P., Jacquin, M. and Choi, D.W. (1999). Calcium ionophores can induce either apoptosis or necrosis in cultured cortical neurons. *Neuroscience*, 90: 1339–1348.

Hackett, P.H. and Roach, R.C. (2000). High-altitude medicine. In *Wilderness Medicine* (eds. P.A. Auerbach), St Louis: Mosby, 1–37.

Hackett, P.H., Rennie, I.D. and Levine, H.D. (1976). The incidence, importance and prophylaxis of acute mountain sickness. *Lancet*, 2: 1149–1154.

Hackett, P.H. (1999). The cerebral etiology of high-altitude cerebral edema and acute mountain sickness. *Wilderness and Environmental Medicine*, 10: 97–109.

Hackett, P.H. and Roach, R.C. (2001). High-altitude illness. *The New England Journal of Medicine*, 345: 107–114.

Halliwell, B. (1999). Antioxidant defence mechanisms: from the beginning to the end (of the beginning). *Free Radical Research*, 31: 261–272.

Halliwell, B. (2001). Role of free radicals in the neurodegenerative diseases: therapeutic implications for antioxidant treatment. *Drugs Aging*, 18: 685–716.

Halliwell, B. (2007). Biochemistry of oxidative stress. Biochemical Society Transactions, 35: 1147-1150.

Halliwell, B., Chirico, S. (1993). Lipid peroxidation: its mechanism, measurement, and significance. *American Journal of Clinical Nutrition* 57(5 Suppl): 715S-724S.

Hemi, T., Ikeda, M., Yasuhara, T. and Murota, S. (2003). Oxidative neuronal death caused by glutamate uptake inhibition in cultured hippocampal neurons. *Journal of Neuroscience Research*, 71: 679–688.

Hopkins, R.O., Kesner, R.P. and Goldstein, M. (1995). Item and order recognition memory in subjects with hypoxic brain injury. *Brain and Cognition*, 27: 180–201.

Hota, S.K., Barhwal, K., Singh, S.B. and Ilavazhagan, G (2007). Differential temporal response of hippocampus, cortex and cerebellum to hypobaric hypoxia: a biochemical approach. *Neurochemistry International*, 51: 384-90.

Hota, S.K., Barhwal, K., Singh, S.B., Sairam, M. and Ilavazhagan, G. (2008). NR1 and GluR2 expression mediates excitotoxicity in chronic hypobaric hypoxia. *Journal of Neuroscience Research*, 86:1142-52.

Hota, S.K., Barhwal, K., Singh, S.B. and Ilavazhagan, G. (2008). Chronic hypobaric hypoxia induced apoptosis in CA1 region of hippocampus: a possible role of NMDAR mediated p75 ^{NTR} upregulation. *Experimental Neurology*, 212:5-13.

Hota, S.K., Barhwal, K., Baitharu, I., Prasad, D., Singh, S.B. and Ilavazhagan, G. (2009). Bacopa monniera leaf extract ameliorates hypobaric hypoxia induced spatial memory impairment. *Neurobiology of Disease*, 34:23- 39.

Sunil, Kumar, Hota., Kalpana, Barhwal, Hota., Dipti, Prasad., Govindasamy, Ilavazhagan and Shashi, Bala, Singh. (2010). Oxidative-stress-induced alterations in Sp factors mediate transcriptional regulation of the NR1 subunit in hippocampus during hypoxia. *Free Radical Biology & Medicine*, 49:178–191

Larson, E.B., Roach, R.C., Schoene, R.B. and Hornbein, T.F. (1982). Acute mountain sickness and acetazolamide. Clinical efficacy and effect on ventilation. *Journal of American Medical Assistance*, 288: 328–332.

Magalhães, J., Ascensão, A., Viscor, G., Soares, J., Oliveira, J., Marques, F. and Duarte, J. (2004) Oxidative stress in humans during and after 4 hours of hypoxia at a simulated altitude of 5500 m. *Aviation and Space Environment Medicine* 75:16-22.

McArdle, W.D., Katch, F.I., and Katch, V.L. (2001). Exercise Physiology: Energy, Nutrition, and Human Performance, Fifth Edition. Lippincott Williams & Wilkins Publishers. Baltimore, MD.

McFarland, R.A. (1937). Pscycho-physiological studies at high altitude in the Andes II: Sensory and motor responses during acclimatization. *Comparative Psychology*, 23: 227–258.

Moore, L.G., Harrison, G.L., McCullough, R.E., McCullough, R.G., Micco, A.J. Tucker, A., Weil, J.V. and Reeves, J.T. (1986). Low acute hypoxic ventilatory response and hypoxic depression in acute altitude sickness. *Journal of Applied Physiology*, 60: 1407–1412.

Moller, P., Loft, S., Lundby, C. and Olsen, N.V. (2001). Acute hypoxia and hypoxic exercise induced DNA strand breaks and oxidative DNA damage in humans. *FASEB* J. 15: 1181 1186.

Muthuraju S., Maiti P., Solanki P., Sharma A.K., Amitabh, Singh S.B., Prasad D., Ilavazhagan G. (2009). Acetylcholinesterase inhibitors enhance cognitive functions in rats following hypobaric hypoxia. *Behavioral Brain Research*, 203 : 1-14.

Pelamatti, G., Pascotto, M., Semenza, C. (2003). Verbal free recall in high altitude: proper names vs. common names. *Cortex*, 39: 97-103.

Phillips, L.W. and Pace, N. (1966). Performance changes at moderately high altitude: short-term memory measured by free recall. *Psychological Reports*, 19: 655–665.

Roach, R.C., Bärtsch, P., Oelz, O. and Hackett, P.H. (1993). The Lake Louise acute mountain sickness scoring system. In: *Hypoxia and molecular medicine* (eds. J.R. Sutton, C.S. Houston, G. Coates), Queen City Printers, Burlington, pp. 272–274.

Robergs, R.A. and Keteyian S.J. (2003). Fundamentals of Exercise Physiology for Fitness, Performance, and Health, Second Edition. McGraw-Hill Higher Education. New York, NY.

Severinghaus, J.W., Chiodi, H., Eger, E.I., Brandstater, B. and Hornbein, T.F. (1966). Cerebral blood flow in man at high altitude. Role of cerebrospinal fluid pH in normalization of flow in chronic hypoxia. *Circulation Research* 19: 274–282.

Singh, I., Khanna, P.K., Srivastava, M.C., Lal, M., Roy, S.B. and Subramanyam, C.S.V. (1969). Acute mountain sickness. *New England Journal of Medicine* 280: 175–184.

Stickney, J.C. and Van, L.E.J. (1953). Acclimatization to low oxygen tension. *Physiological Reviews*, 33: 13–34.

Titus, A.D., Shankaranarayana, Rao, B.S., Harsha, H.N., Ramkumar, K., Srikumar, B.N., Singh, S.B., Chattarji, S. and Raju, T.R. (2006). Hypobaric hypoxia-induced dendritic atrophy of hippocampal neurons is associated with cognitive impairment in adult rats. *Neuroscience*, 145: 265–278.

Topor, Z.L., Pawlicki, M. and Remmers, J.E. (2004). A computational model of the human respiratory control system: responses to hypoxia and hypercapnia. *Annals of Biomedical Enginering*, 32: 1530–1545.

West, J.B. (1986). Hypoxia of altitude. *Lancet*, 2: 1097.

West, J.B. (1996). Prediction of barometric pressures at high altitude with the use of model atmospheres. *Journal of Applied Physiology*, 81: 1850–1854.

Williams A., Health D., Kay J.M., Smith P. (1977). Lung mast cells in rats exposed to acute hypoxia, and chronic hypoxia with recovery. *Thorax*. 32: 287-295.

Xie, A., Skatrud, J.B., Khayat, R., Dempsey, J.A., Morgan, B. and Russell, D. (2005). Cerebrovascular response to carbon dioxide in patients with congestive heart failure. *American Journal of Respiratory Critical Care and Med*icine 172: 371–378.

□□□

Chapter-22

Agro-Bioterrorism : A Potential Threat and Preparedness

Gyan P Mishra, Raghwendra Singh[1], Manish Bhoyar, Jitendra Kumar, Narendra Singh and RB Srivastava

Defence Institute of High Altitude Research, Defence R&D Organisation, C/o 56 APO, Leh, Ladakh, J&K, India, 901 205

[1]Directorate of Weed Science Research, Jabalpur, Madhya Pradesh, 482 004

1. Introduction

Destruction of food crops or economy sustaining crops of the enemy may be termed as agri-biowarfare or in case of terrorism agri-bioterrorism. The purpose behind this kind of attack is to increase the hunger of the population and subject them to humiliation, death and disease. Manipulation of food supply or food crops by other means is also a form of bio-warfare with far reaching consequences (Sen and Omnibus, 1999). The total loss in crop production resulting from its main enemies, viz. diseases and insect pests, often reaches staggering proportions all over the world (Table 1).

Table 1 : Percentage of crop produce lost because of diseases and Insect pests per year (Oerke *et al.*, 1994).

Continent/Region	Loss (%)	Loss (in billion of USA dollars)
North and Central America	31.2	23.0
Europe	28.2	16.8
Russia	40.9	22.9
South America	41.3	21.8
Africa	48.9	12.8
Asia	47.1	145.3
*India (Dhaliwal *et al.*, 2004)	20.4	689400 million rupees

A number of plant pathogenic microbes and insect pests which caused sever losses in different countries are presented in Table 2 and 3 respectively. These insects and pests can be considered as potential bio-warfare agents.

Table 2 : Some examples of severe losses caused by plant diseases

Diseases	Locations	Remarks
Fungal Diseases		
Brown spot of Rice (*Helminthosporium oryzae*)	Asia	**Epidemics** - Great Bengal Famine (1942)
Late blight of Potato (*Phytophthora infestans*)	All over the world	**Epidemics** - Irish Famine (1845-1846)
Downey mildew of grape (Plasmopaara viticola)	Europe, U.S.A.	European **epidemic** (1880)
Coffee rust (Hemileia vastatrix)	Sri Lanka, Brazil, South and central American and African states	Severe **epidemics** (1870s to 1880s) totally destroyed coffee plantation in Sri Lanka
Southern Corn leaf blight (*Helminthosporium maydis* race T)	U.S.A	**Epidemic** -1970
Cereal rusts, particularly Wheat rusts (*Puccinia* spp.)	World wide	Continued losses, occasionally very heavy
Wheat smuts (*Ustilago* spp.)	World wide	Continued losses, occasionally very heavy
Sigatoka disease of banana (*Mycosphaerella musicola*)	World wide (particularly in West Indies and Asia)	Continued losses, occasionally very heavy
Bacterial Diseases		
Angular leaf spot of Cotton (*Pseudomonas malvacearum*)	U.S.A, Mexico	Heavy losses annually
Fire Blight (*Bacillus amylovorus*)	North America, NewZealand,	Continued losses, occasionally very heavy

contd....

Bacterial leaf Blight (*Xanthomonas oryzae*)	Japan, China, South east Asia	Heavy losses annually
Citrus canker (*Xanthomonas compestris* var. citri.)	World wide	Severe damage to plantations in Florida (U.S.A) in 1910s and 1980s
Viral Diseases		
Sugarbeet yellows	World wide	Heavy losses annually
Sugarcane mosaic	World wide	Heavy losses
Potato Virus	World wide	Heavy losses
Citrus tristeza	America, Africa	Millions of trees killed
Nematode Diseases		
Root Knot (mostly of vegetables: *Meloidogyne* spp.)	World wide	Substantial annual losses
Ufra Diseases (*Ditylenchus angustus*)	Bengal, Burma, Malaya	Heavy damage to crop

Table 3 : Some examples of severe losses caused by Insect Pests (http// www.ncbi.nlm.nih.gov)

Diseases	Remark
Chir pine defoliator (*Crytothelia cramerii*)	First **epidemic** reported in 1885 from Tons Valley, Uttranchal. Subsequently recorded from H.P. (1928), Kahhula, Pakistan (1934). Recently reported from Rajouri (J&K) in 1989 – 1990. 5% mortality in 2000 ha. area, with 0.3 million trees in J&K; net loss 22.5 million rupees.
Sal heart wood borer (*Hoplocerambyx spinicornis*)	**Epidemic** dates back to 1899 in Singhbhoom, Bihar. Reported from Assam (1906, 1961), H.P. (1'948 – 1952), M.P. (1905, 1927 – 28, 1948-52, 1959-63), Uttranchal (1916-24, 1934-37, 1958-60, 1961, 1965), West Bengal (1931-34). Recently a very heavy epidemic occurred in M.P. in 1998, affected some million sal trees.
Toon shoot borer (*Hypsipyla robusta*)	A serious pest of toon and mahogony, capable of causing 100% mortality in seedlings and young plantations. In India, some of the seriously infested toon plantations were destroyed, causing loss of R.15-30 per acre. Also reported to cause damage in Sri Lanka, Australia, Bangla Desh, Pakistan, Nigeria and West Indies).

contd....

Teak skeletonizer (*Eutectona machaeralis*)	Major pests of teak, occurring throughout south Asia and some parts of South-East Asia. The studies carried out in the past estimate the loss to about 0.051 millions/ha/year.
Shisham defoliator (*Plecoptera reflexa* Guenee)	Serious **epidemic** in Changa Manga and Khanewal forest divisions (now in Pakistan) in 1899. Serious epidemics have been recorded from Chichawatni and Khanewal in 1927, 1928, 1932 and in Ambala forest division in 1974 and 1975.
Khasi pine defoliator (*Eterusia pulchela*)	A large scale **epidemic** occurred in 1975 in 7500 ha. of Jaintia hills and 2500 ha. in Khasi hills. Affected stands of 5-30 years; heavy mortality (50%). Heavy defoliation occurred again in 1978. Two or more complete defoliations are sufficient to kill the tree.
Poplar defoliator (*Melosoma populi*)	A serious pest of Poplars and Willows in the temperate Himalayas from J&K to Arunachal Pradesh.
Teak defoliator (*Hyblea puera Gram*)	Pest **epidemics** reported from time to time.
Desert locust (*Schistocerca gregaria*)	Damage to crop caused by locusts during 1926 -31 cycle was estimated at 100 million rupees in India (Atwal and Dhaliwal, 2005)
Cotton Bollworm (*Helicoverpa armigera*)	A serious pest of cotton and pulses. Reported in Punjab, Haryana, Andhra Pradesh and Gujarat. Heavy losses annually.

2. Preparedness to Face Bioterrorism

The degree of damage an organism or a toxin can inflict, varies widely and this also depends upon the socio-economic status of the country or sections of its vulnerable population. Some aspects of preparedness against bioterrorism are similar to those needed to face natural disasters like cyclones, floods and earthquakes, as plant epidemics often follow such disasters. We should act in advance and not after the disaster strikes, as we usually do. Some urgent measures in this regard are:

a. *Public awareness*: It is reasonable to expect that the bioterrorist would choose a particular disease for use in a particular country, and may even target a particular segment of the population of that country. For example, a plant pathogen can be transmitted through susceptible or carrier to the crops. The crop can be affected by the New disease followed by an epidemic which may cause huge losses. Public should also be made aware of the precautionary measures to be taken and the methods of management of the diseases, when they appear. Facts about different diseases of the nature of bioterrorist import should be publicised, the way Centres for Disease Control and Prevention in US do. All this is not a small task in countries with large illiterate populations and an infrastructure to achieve this must be built up, without loss of time.

b. *Research and Development*: We need adequate number of well equipped labs with facilities for microbiologists to develop quick and certain means to identify the pathogen and the disease, for biochemists to develop diagnostic kits, for pharmacologists to develop drugs. In every country there is some activity of this nature but it is not adequate.

c. *Contingency plan of action*: Authorities should draw contingency plans of action, separately for each vulnerable area, to meet with the situation when it develops, and identify the disease, etiology and control measures to face the situation.

3. Potential Diseases which can be Used as Bio-Warfare Agents

There are many potential diseases which can be used as bio-weapons for large scale crop destruction. The details viz. name of disease, symptoms, agent, occurrence and climatic factors, incubation period, mode of transmission and prevention control of some of them are given below (Table 4).

4. Potential Insect Pests which can be used as Bio-Warfare Agents

There are many potential insect pests which can be used as bio-weapons for large scale crop destruction. The details viz. name of insect-pest, symptoms, agent, damaging stage and prevention control of some of them are given below (Table 5).

5. Locust Outbreak in Ladakh in the Year 2006 by *Locusta migratotria*

According to J&K Govt. Deptt. Of Agriculture, an area of 6,200 ha in Ladakh has been affected and destroyed by the locusts, of which 5,000 ha in Changthang (Nyoma block, Leh Distt.) and 1,200 ha in Zanskar sub-division (Kargil Distt.) (Fig. 1).

Table 4 : Potential diseases which can be used as bio-warfare agents

Name of Disease	Symptoms	Agent	Occurrence & Climatic factors	Incubation Period	Mode of transmission	Prevention & Control
Late blight of Potato (Potato)	Necrois in lower leaves	*Phytophthora infestans*	Temp. 10-20^{0}C Humidity above 75%, frequent rain and cloudy weather	3-4 days	Air borne	Fungicide: Mancozeb and Dithio carbamate 1.7-2.0 kg/ha
Brown Spot (Rice)	Small brown oval or oblong spots on leaf blades and leaf sheath	*Helminthosporium oryzae*	Low temp. 18-20^{0}C	5-6 days	Seed borne Secondary infection by Airborne through conidia	Pre soaking in cold water and then hot water treatment Fungicide: Organo mercury seed dressing substances such as ethyl mercury phosphate, ethyl mercury chloride
Rust (Wheat)	Uridias are circular, oblong, circular orange red, brown or powdery	*Puccinia graminis tritici, Recondita striiformis*	Low temp. and wind blowing from north to south	5 -10 days	Air borne	Development of resistant varieties

Downey mildew (Grapes)	Cottanic growth at lower surface of leaves	*Plasmopora viticola*	20-25^{0}C	5-18 days	Airborne	Apply Bordeaux mixture and captan, Development of resistant varieties
Angular leaf spot of cotton	Leaf spots on the leaves, black lesions on the stem, sunken lesions on the bolls	*Pseudomonas malvacearum*	Soil temp. 30-40^{0}C and Humidity 85%	3-10 days	Soil borne	Seed treatment with frequent stirring in Con. H_2SO_4 for5 to 15 min.
Fire Blight (Apple)	Blossom blight, leaf blight, Twig blight, fruit spot blight	*Bacillus amylovorus*	Rain followed by warm cloudy weather	5-6 days	Aphids	Apply Bordeaux mixture, complete removal of small branches bearing lesions and expose surface protected by asphalt paint and winter pruning
Bacterial leaf Blight of Rice	Water soaked spots are formed along the margin of leaf blade or along the prominent veins	*Xanthomonas oryzae*	Combination of rainy weather, dull windy weather and temp. of 20-26^{0}C	-	Soil borne and through infected grains	Spraying of Copper fungicide and Developing resistant varieties
Ufra disease (Rice)	Chlorotic streaks develop in the upper leaves	*Ditylenchus angustus*	16-19^{0}C	-	Soil borne	Burning stubble & straw, use of early maturing varieties & rotation with jute

Table 5 : Potential insect pests which can be used as bio-warfare agents

Name of Insect Pest	Symptoms	Agent	Damaging Stage	Prevention & Control
American boll warm (cotton)	Larvae or polyphagus they feed on flower parts and foliage	*Helicoverpa armigera*	Larvae	Apply NPV @ 250-500 larval equivalents / ha, 2.5 lts of endosulfan 35EC, use of Bt Cotton transgenic varities
White Grub (Fruit trees)	Nymph feed on rootlets resulting complete withering and drying up of seedlings and young plants	*Holotrichia incularis*	Nymph	Lindane 2% dust @ 50kg / ha
Pyrilla (sugarcane)	Leaf turns pale yellow and shrivel up later	*Pyrilla perpusilla*	Nymph and adult	Remove sprouts and stubble from field Spray 1-1.25 lts endosulfan 35EC/ha in 375 lts of water
Stem borer (Rice)	Ears devoid of grain	*Scirpophaga incertulus*	Caterpillar	Use of resistant varieties viz. *Ratna, Sasyasree, Saket*, Removal and destruction of stubbles, 10kg of Chlor pyriphos 10G /ha in standing water

Army worm (Wheat)	Whole leaves including the mid rib are consumed and the field looks as if grazed by cattle	*Mythimma separate*	Caterpillar	Spray 3kg of carboryl 50WB / ha
Mango hoppers (Mango)	Inflorescence weathered and turned to brown, sooty mould develops on patches of honey dew exuded by the nymph	*Amritodus atkinsoni*	Nymph and adult	Spraying with malathion LVC @ 1.4 lts / ha
Rhinoceros beetle (Coconut)	The injuries are seen as a series of holes on the fronts. The growing point is soon cut off and tree dies	*Oryctes rhinoceros*	Grub	Hook out the beetles from the affected crown by means of hooks during July-August Treat the breeding places with 2.0kg of Carbaryl50WB in 250 lts of water/ha
Wooly Apple Aphid (Apple)	Swelling and knots appear on the roots	*Eriosoma lanigerum*	Nymph	Use of resistant root stock like Golden Delicious, Morton Stocks 778, 779, 789 and 793. release exotic parasite, *A. mali*. Fumigation with paradichlorobenzene granules.
Desert locust	Consume all vegetation and cause a famine	*Schistocerca gregaria*	Hoppers and adults	The adults can be beaten to death with thorny sticks, Dusting of Lindane 2%, Destruction of nymphs

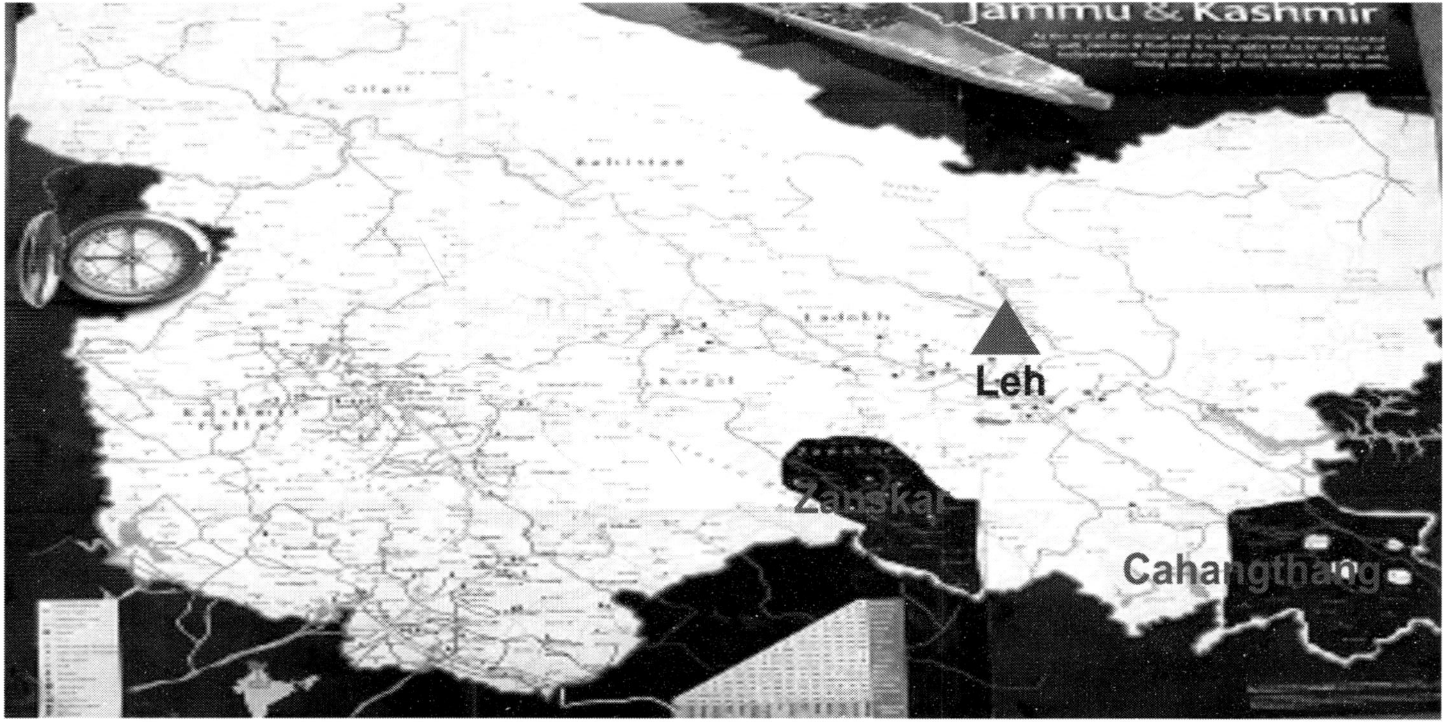

Area affected by locust

Villagers not allowed to spray insecticides Viz. Demchok, Skakjung, Koyul and Rhongo

Fig. 1 : Map of Jammu and Kashmir Showing Locust affected areas:

6. Drug Materials for Controlling Epidemic Diseases

a. *Fungicides*: Vitavax, Carbendazim (Bavistin), Benomyl (Benlite), Dithane M- 45, Dithane Z-38, Tilt 25 EC, Karathane, Ethrimol, Diethrimol, Calixin, Thiram, Agrosan GM, Ceresan, Captan, Copper Oxi chloride, Zineb, Ziram, Emisam, Metalaxyl, Carboxin, Captafol, Aldicarb, Propionic acid, Sodium meta bi suphite etc.

b. *Antibiotics*: Streptomycin, Terramycin, Neomycin, Kasugamicin, Hygromycin

c. Insecticides

   c.1 Botanical Insecticides : Nicotin sulphate , Pyrethrum, Rotenone

   c.2 Chlorinated Hydrocarbons: Aldrin, BHC, Endosulfan, Lindane, DDT, Dicofol, Endrin, Ethylene dibromide, Heptachlor

   c.3 Organophosphate: Bromophos, Diaxinon, Demeton, Malathion, Monocrotophos, Parathion, Menazon, Phorate

   c.4 Carbamates: Aldicarb, Carbaryal, Carbofuran

   c.5 Synthetic Pyrethroids: Permethrin, Cypermethrin, Decamethrin

7. Emerging Drug Materials

In countries like India and China, the indigenous systems of medicine have a number of plant based drugs effective against several infectious diseases that can be a means of bioterrorism. The advantages of indigenous systems are accessibility, economic viability and reasonable effectiveness, in terms of both prevention and management of infectious diseases, with a local import.

Collectively, plants produce a remarkably diverse array of over 100,000 low molecular mass natural products, also known as secondary metabolites. Secondary metabolites are distinct from the comments of intermediary metabolism in that they are generally non essential for the basic metabolic processes of the plant (Dixon, 2001). Most are derived from the isoprenoid, phenylpropanoid, alkaloid or fatty acid/ polyketide pathways. This rich diversity results in part from an evolutionary process driven by selection for acquisition of improved defence against microbial attack or insect/ animal predation. However such diversity has made it difficult to apply conventional molecular and genetic techniques to address the functions of natural products in plant defence, or to improve plant disease resistance using metabolic pathway engineering (Pichersky and Gang, 2000). Related plant families generally make use of related chemical structures for defence (for example, isoflavonoids in the Leguminosae, sesquiterpenes in the solanaceae), although some chemicals classes are used for defensive functions across taxa (for example, phenylpropanoid derivatives). Some species produce a broad range of antimicrobial compounds. For example, Cocoa, when infected by the vascular wilt fungus *Verticillium dahilae*, accumulates the pentacyclic triterpene arjunolic acid, two hydroxylated acetophenones and, most unusually, elemental sulphur, the only known inorganic antimicrobial agent produced by plants (Cooper, 1996).

a. Plant Sources of Natural Drugs

Research work is in progress to exploit the potential of plant extracts and antagonistic micro organisms for control of plant diseases. This method of disease management is devoid of any health hazard problems. Oil extracted from *Capillipedium foetidium* exhibited activity against *Aspergillus niger, A. flavus*. Control of hill bunt of Wheat (*Tilletia foetida*) was obtained through seed treatment with extracts of species of Datura, Thuja, and Eucalyptus. Besides this many other plant extracts to control various pathogens are presented below (Table 6).

Table 6 : Plant extracts to control various pathogens

Plant Species	Pathogen controlled	Reference
Allium cepa, Allium sativa, Azadirachta India, Datura stramonium, Ocimum sanctum	*Alternaria brssicola, Colletrichum capsici, Fusarium oxysporum*	Shivpuri *et al*, 1997
Vinca rosea, Lantana camara, Ocimum tenuiflorum, Azadirachta India	*Pyricularia oryzae, Helminthosporium oryzae*	Ganguly, 1994
Acasia auriculiformis, Ricinus communis, Zinger officinalis, Tamarindus indica, Lawsonia inermis	*Fusarium moniliformae*	Gohil and Vala, 1996
Vitex negundo, Catharanthus rosea, Aloe vera	Rice tungro	Selvaraj and Narayanasamy, 1991

b. Bio-control agents effective against plant pathogens

Currently pesticide registrations are lost due to concerns about their safety and environmental impacts. Further, pesticides are high energy inputs and are also costly. Biological control has thus become a critical component of integrated disease management. A large number of formulations of biocontrol microorganisms effective particularly against soil-borne plant pathogens have been reported since 1980 (Lewis, 1991; Table 7).

Table 7 : Commercial formulations with Bio- control agents effective against plant pathogens (Lewis, 1991).

Target pathogen	Antagonist	Name	Source
Chondrostereum purpureum	*Trichoderma sp.*	BINAB-T	Bio-InnovationAB, Toreboda, Sweeden
Pythium ultimum/ Cotton	*Pseudomonas fluorescens*	Dagger G	Ecogen, Inc. Langhorne, PA
Agrobacterium tumefaciens/ Crucifers	*Agrobacterium radiobacter*	Galltrol	Ag Bio Chem, Inc. Orinda, CA. USA
Alternaria brassicicola/Trees	*Streptomyces*	Mycostop	Kemira Oy. Vass, Finland
Agrobacterium tumefaciens/Trees	*Agrobacterium radiobacter*	Norbe 84-C	New Bioproducts, Inc. Corvallis, CA, USA
Heterobasidion annosum/Trees	*Phlebia gigantea*	P.g. Suspension	Ecological Labs. Ltd., Dover, UK

8. Geneticially Modified Organisms (GMOs)

One of humankind's greatest endeavour's has been to modify plant species to meet our domestic and economic needs. Recent advances in plant molecular biology have opened new avenues for the production of genetically engineered plants and in the precise transfer of novel genes into crop plants from diverse sources, not feasible before (Jauhar, 2001). The majority of trails were designed to test for enhanced resistance to viruses, bacteria, fungi, insects, herbicides and diseases. GMOs are a part of regular agricultural practice. Although they are just being introduced in agriculture in some parts of the world, they are extensively used elsewhere. In the USA, 40 million acres (16 million hectares) are said to be under genetically modified crops, including cotton. GMOs are very much a reality and will certainly be the future of agriculture. The development of pest-resistance Bt crops (corn, soybean and cotton), introduction of virus coat protein-mediated resistance to viral diseases, and enhancing nutritional level of crop plants are some of the accomplishments of transgenic technology (Ye *et al.*, 2000). Certain pests such as European corn borer (ECB) cause a loss of one to two billion dollars per year in the United States (Hyde *et al.*, 1999). However, a gene from a natural soil bacterium, *Bacillus thuringiensis* (Bt), when bioengineered into the corn genome, confers resistance to ECB, largely eliminating damage from this serious pest.

Another major breakthrough in providing built-in genetic resistance to diseases is the discovery that expression of a virus coat protein gene as a transgenic in a plant confers resistance to that virus in direct proportion to the amount of coat protein produced by the transformed plant (Beachy *et al.*, 1990). This novel approach has opened up new avenues of control of viral diseases in plants. Thus, papaya ring spot in Hawaii has been controlled by coat-protein mediated resistance (Gonsalves, 1998). Transgenic approaches have also been employed to combat fungal diseases in cereal crops including wheat and barley (Dahleen *et al.*, 2001: Muthukrishan *et al.*, 2001). Production of the first transgenic durum wheat by microprojectile bombardment and standardization of the technology of durum transformation (Bommineni *et al.*, 1997) opened up avenues of engineering with antifungal genes. The biotechnological tools are used to evolve genetically modified plants against serious diseases and insect pests which can cause epidemic. Some transgenic plants carrying genes for insect resistance are presented as under (Table 8).

9. Bio-Safety Precuations

Bio-safety is a strategic and integrated approach that encompasses the policy and regulatory frameworks (including instruments and activities) that analyze and manage risks in the sectors of food safety, animal life and health, and plant life and health, including associated environmental risk. Bio-security covers the introduction of plant pest's, animal pests and diseases, and zoonoes, the introduction and release of genetically modified organisms (GMOs) and their products, and the introduction and management of invasive species and genotypes. Bio-safety is a holistic concept of direct relevance to the sustainability of agriculture, food safety, and the protection of the environment, including biodiversity.

Table 8 : Some transgenic plants carrying genes for insect resistance. (Khush and Brar, 1999)

Crop	Foregin gene	Origion of gene	Target insect pest (s)
Brinjal	cry (iii) b	*Bacillus thuringiensis*	*Leptinotarsa decemlineata*
Cotton	cry 1A (b), cry 1A (c)	*B. thuringiensis*	*Helicoverpa zea, Spodoptera exigua, Trichoplusia ni*
Maize	cry 1A (b)	*B. thuringiensis*	*Ostrinia nubilalis*
Potato	cry 1A (b), Oryza cystatin 1 (oc 1)	*B. thuringiensis Rice*	*Phthorimaea operculella L. decemlineata*
Rice	cry 1A (b)	*B. thuringiensis*	*Scirpophaga incertulas*
Sugarcane	cry 1A (b)	*B. thuringiensis*	*Diattraea sachharalis*
Tomato	Cry 1A (c), Bt (k)	*B. thuringiensis B. thuringiensis*	*Malacosoma sexta Heliothis zea*

Why do we need bio safety?

With the growing threats of terrorism and the gradual increase in Global Warming, there are many things threatening our Agriculture. Terrorism brings the threat of bio-terriosm, which is a form of terrorism used to target farms to disrupt the food supply. The global warming is also giving suitable habitat for natural outbreaks to occur.

Steps for establishing an effective and complete bio-safety

1. Post emergency response phone numbers and contact list (phone numbers for suspected bio-safety problems, distributors of chemicals, pesticides and utility companies
2. Complete a risk assessment checklist: It is a tool for quickly identifying potential hazards and risks that increase producer vulnerability to vandalism, theft, bio-safety problems and other emergency situations. This check list does not encompass every imaginable bi-safety risk, but it does delineate some of the bio-safety hazards and concerns that may be overlooked by those in the agricultural sector.
3. Create maps and records of your operation: Aerial photographs should also be used, topo maps also should be included

4. Enhance your crop management and your pest identification skills.
5. *Sponsor an Emergency Responders' Tour of Operation and Training Event:* This will allow for emergency crews to be familiar with the operation in the event of an emergency. Also the training event will allow for quick response and control if an emergency was to occur.
6. Evaluate, revise, and update your preparedness plan and involve others in preparedness planning.

Conclusions

The result of an Agro-warfare attack is an epidemic or an epizootic or increase of hunger due to crop failure. To combat any attack, a sound health infrastructure is a must. At the outset, rapid diagnosis of the disease, communication of information and identification of the organism is absolutely necessary. If eradication of the pest or disease is not possible, an endemic infestation would result in a lower equilibrium level of production or quality. Long-term economic recovery includes resuming the husbandry of animals and plants in the affected areas, introducing new genetic traits that may be necessary in response to the pest or disease, rebuilding confidence in domestic markets, and regaining international market share. Resources would be devoted to acquiring plant varieties with resistance characteristics and breeds of animals more suitable to the new environment. The social sciences and public health institutions play a complementary role to the agricultural sciences in responding to and recovering from agro-bioterrorism. The most important task of the administration is to curtail the panic and mass hysteria of the public after the outbreak of epidemic.

References

Atwal, A.S. and Dhaliwal, G.S. 2005. Agricultural pests of South Asia and their management. Kalyani Publishers. New Delhi. Pp. 366.

Beachy, R.N., Loesch-Fries, S. and Turner, N.E. 1990. Coat protein mediated resistance against virus infection. *Ann. Rev. Phytopathol.* 28: 451-474.

Bommineni, V. R., Jauhar, P.P. and Peterson, T.S. 1997. Transgenic durum wheat by microprojectile bombardment of isolated scutella. J. Hered. 88. 475-481.

Cooper, R.M. 1996. Detection and cellular localization of elemental sulphur in disease resistant genotypes of Teobroma caco. *Nature*, 379, 159-162 .

Dahleen, L.S., Okubara, P.A. and Blechl, A.E. 2001. Transgenic approaches to combat Fusarium head blight in Wheat and barely. *Crop Sci.* 41.

Dhaliwal, G.S., Arora, R. and Dhawan, A.K., 2004. Crop losses due to insect pests in Indian Agriculture: an update. *Indian J. Ecol.* 31(1): 1-7.

Dixon, R.A. 2001. Natural products and plant disease resistance, *Nature.* Vol. 411. 14 June.

Ganguly, L.K. 1994. Fungitoxic effect of certain plant extracts against rice blast and brown spot pathogen. *Enviroment and Ecology.* 12(3): 731-733.

Gohil, V.P. and Vala, D.G. 1996. Effect of extracts of some medicinal plants on the growth of Fusarium moniliformae. *Indian J. Mycol. Pl.Pathol.* 26(1): 110-111.

Gonsalves, D. 1998. *Control of ring spot virus in papaya: A case study. Ann.* Rev. Phytopathol. 36: 412-437.

http//www.ncbi.nlm.nih.gov

Hyde, J., Martin, M.A., Preckel, P.V. and Edwards, C.R. 1999. The economics of Bt corn: Valuing protection from the European coren borer. *Rev.Agric.Econ.* 21: 442-454.

Jauhar, P.P. 2001. Genetic engineering and accelerated plant improvement: Opportunities and challenges. *Plant Cell, Tissue and Organ Culture,* 64: 87-91.

Khush, G.S. and Brar, D.S. 1999. Transgenic plants for pest management. Progress and future outlook. In: GS Dhaliwal & Ramesh Arora (Eds) Environmental stress in crop plants. Commonwealth publishers, New Delhi. Pp.281-307.

Lewis, J.A. 1991. Formulation and delivery systems of biocontrol agents with emphasis on fungi. : In the Rhizosphere and plant growth. Kluvier Academic Publishers, Netherlands, pp. 279-287.

Muthukrishan, S., Liang, G.H., Trick, H.N. and Gill, B.S. 2001. Enhancement of disease and insect resistance in cereals by genetic transformation with PR-protein genes. *Appl. Genet.*

Oerke, E.C., Dehre, H.W., Schonbeck, F. and Weber, A. 1994. Crop production and crop protection: estimated losses in major food and cash crops. Elsevier. Amsterdam.

Pichersky, E. and Gang, D.R. 2000. Genetics and biochemistry of secondary metabolites in plants: an evolutionary perspective. *Trends Plants Sci.* 5, 439-445.

Selvaraj, C. and Narayanasamy, P. 1991. Effect of plant extracts in controlling rice tungro. *Int. Rice Research Newsletter.* 16(2): 21-22.

Sen, A. and Omnibus, J.D. 1999. The great Bengal Famine. Chap.6. Oxford University Press.

Shivpuri, A., Sharma, O.P. and Shamaria. S.L. 1997. Fungitoxic properties of plant extracts against pathogenic fungi. *J. Mycol. Pl. Pathol.* 27(1): 29-31

Ye, X., Al-Babill, S., Klotl, A., Zhang, J., Lucca, P., Beyer, P. and Potrykus, I. 2000. Engineering the provitamin A biosynthetic pathway in to rice endosperm. *Science*. 287: 303-30.

□□□

Chapter-23

Development of Radiation Countermeasure Agents through Harnessing High-Altitude Cold Arid Region Bioresource

Rajesh Arora and W. Selvamurthy
Office of Chief Controller R&D (Life Sciences and International Collaboration), Ministry of Defence, Govt. of India, DRDO Headquarters, DRDO Bhawan, Rajaji Marg, New Delhi-05

1. Introduction

Weapons of mass destruction (WMD) are weapons that can kill and cause substantial harm to a large section of population and other life forms, besides inflicting significant damage to man-made structures e.g. buildings, natural structures and the biosphere. The term WMD has often been used in a variety of ways, mostly politically rather than technically. Initially, the term was coined with reference to aerial bombing with chemical explosives, but now the term has become more broad and encompasses the use of chemical, biological, radiological or nuclear (CBRN) weapons. Weapons of mass destruction pose a challenge in today's scenario in view of the chances of their use by hostile and non-state actors. A fairly lesser degree of control over such weapons in some countries by way of Government machinery, and the passage of such weapons into less responsible hands via the terror network, which has spread its wings round the globe, poses grave problems for certain nations specifically and the whole of humanity in general. The threat perception is further compounded by activities of hostile countries, non-state actors, terrorists *etc.*, and may be in the form of low-intensity conflicts, tempering/sabotage etc.

Ever since the use of atomic weapons over Japan, threat has loomed large over humanity - what if such WMDs find use in future wars and conflicts? The 'cold war' days are reminiscent of the importance USA and USSR attached to such weapons. The post-USSR political environment is much eased, never-the-less the threat from nuclear, biological and chemical (NBC) agents remains and consequently the military of even smaller nations is gearing up to face this emerging challenge. Strategists fear that the consequence of Chemical, Biological, Radiological, and Nuclear (CBRN) emergencies, arising out of the use of toxic and hazardous chemicals/materials, biological agents and radiological/nuclear materials, will be far more damaging. CBRN emergencies can be caused either by ignorance, negligence, incompetence, accidentally or by way of malevolent intent. The emergence of state-sponsored terrorism, proliferation of chemical/biological weapons, availability of materials and scientific weapons expertise points towards the increasing probability of occurrence of mass casualties resulting due the use of CBRN agents.

2. The Nuclear Threat

Of the chemical, biological, radiological/nuclear (CBRN) agents, the latter i.e., radiological and nuclear agents pose immense concern since their effects can be far reaching, often extending to generations. Though radiological/nuclear sources are heavily guarded and secured by the law, there is a distinct probability in today's scenario of the use of radiological dispersal device (RDD) or radiation exposure device (RED) by terrorists or some rogue nations for gaining tactical advantage or creating mass panic. The threat from Dirty bombs to kill or incapacitate the military/paramilitary forces and the undefended civilian targets is real today. In view of this, it is imperative to prepare in whichever way possible for such untoward eventualities. The ever-increasing use of nuclear power by a large number of nations has made it inevitable for researchers to develop radioprotective drugs that can protect against the severely damaging effects of ionizing radiation. Most countries are quite disturbed by these recent developments since a nuclear eventuality can result in large-scale casualties, besides inflicting immense damage on population as well as the ecosystem. The damaging effects can last for several years in the absence of proper clean-up, which is difficult to achieve in such scenarios. The recent Fukushima incident in Japan, following the massive earthquake and the subsequent Tsunami, led to damage to the nuclear reactors in Japan and is a stark reminder of the threat from such emergencies and has also left several questions regarding the safety of nuclear reactors and the possibility of leakage of radiation into the environment (Sharma and Arora, 2011). It has become absolutely necessary today to develop and stockpile drugs that can counter the deleterious effects of ionizing radiation. Such drugs could be used during radiation emergencies to protect the first responders, rescue personnel and the general population at large.

3. How the killer Ionizing Radiation Inflicts Damage in Biological Systems?

Ionizing radiation causes damage to biological systems via direct and indirect effect. Protecting living organisms against the initial transfer of radiation energy due to the extremely rapid transfer of energy that usually takes place in less than a billionth of a second is a difficult enterprise. However, the damage due to radiation injury resulting due to low-LET radiation is modifiable. Exposure of biological systems to ionizing radiation results in the formation of reactive oxygen species (ROS) and reactive nitrogen species (RNS). These reactive species inflict damage to the various bio-macromolecules like DNA, lipids and proteins (Fig. 1), (Prasad, 1999; Sharma and Arora, 2006) leading to induction of apoptosis and consequently death.

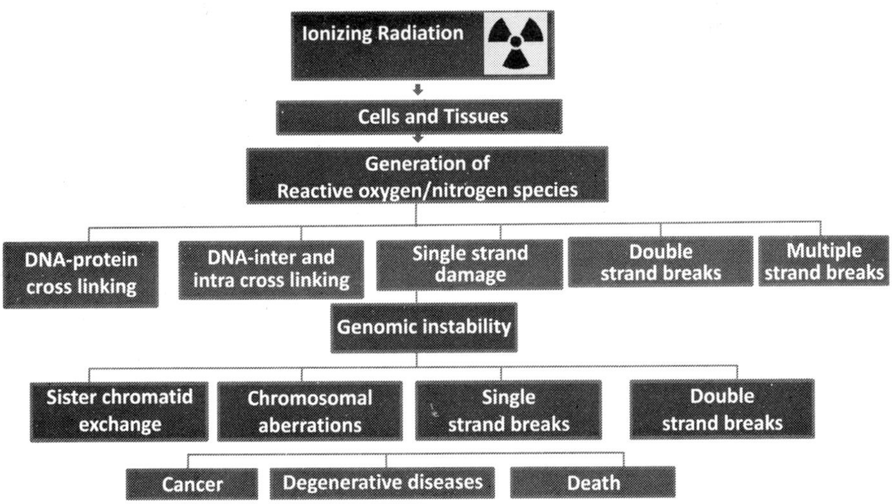

Fig. 1 : Damage caused by ionizing radiation to DNA and its consequences.

The health effects resulting from exposure to ionizing radiation are classified into two categories: stochastic (probabilistic) and deterministic. Probabilistic effects can take several years to develop (e.g., cancer appearing several years later after exposure), while deterministic effects (e.g., cataract induction, hematologic deficiencies, skin erythema, damage to skin and fertility impairment) at high radiation doses are manifested with certainty above a threshold. Individuals exposed to radiation may develop radiation sickness or exhibit signs of radiation injury. Radiation sickness develops as a result of the acute or delayed consequences of exposure of total body to very high doses of ionizing radiation resulting in non-specific clinical symptoms and hematological alterations. On the other hand, radiation injury refers to the acute or delayed consequences of exposure of a small part of the body to high doses of ionizing radiation resulting in severe burns, induction of cataract, pneumonia and hypothyreosis (Milosevic *et al.*, 1988; Metller and Voclz, 2002).

The dream of radiation biologists is of radiation countermeasure agents that are highly efficacious, possessing a wide margin of safety and tolerance in living systems, are stable for several years and easy to dispense and cost-effective and above all would be a panacea for radiation sickness (Fig. 2).

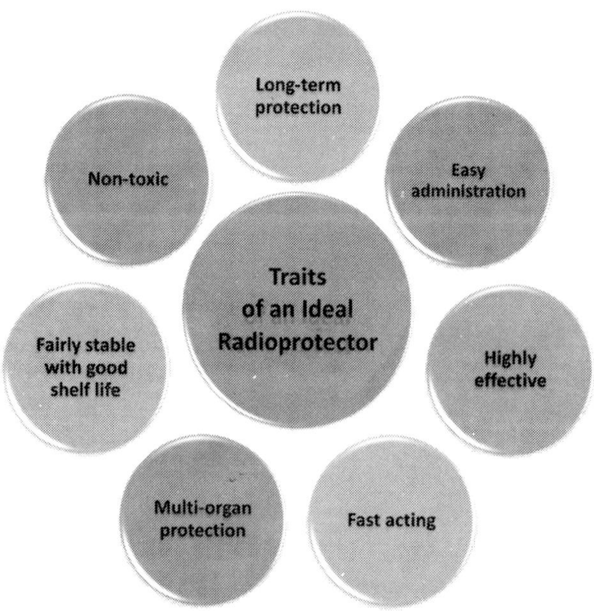

Fig. 2 : Traits of an ideal radioprotector

The use of radioprotectants in a nuclear theatre of operations is going to be indispensable, it is very important that effective human (military) performance levels are maintained after their administration (Giambaressi and Jacobs, 1987). To date, there is not a single radioprotective agent available for human application that meets all the requisites of an ideal radioprotector i.e., possesses no toxicity, offers long-term protection, is effective over a wide range of high/low temperature, possesses a stable shelf life of several years, and can be easily administered, preferably *via* oral route (Maisin, 1998; Nair *et al.*, 2001; Coleman *et al.*, 2003; Weiss and Landeur, 2003; Arora, 2008; Arora *et al.*, 2005a, 2006, 2007a, 2008a, 2010a). The quest for the ideal radiation countermeasure agent is continuing and it may take several years before we have such an agent. Since radioprotection is a multifaceted phenomenon, it is unlikely that a single compound will have the ability to protect all target organs and systems and most likely will be a cocktail of bioactive molecules, many of which would be derived from nature i.e., would be natural or nature-identical molecules.

4. The Quest for Radiation Countermeasure Agents

While the quest for development of radiation countermeasure agents is underway, several researchers have attempted to classify radioprotectors into different categories and there is no universal classification that is unanimously acceptable. However, currently the best classification available defines a radioprotector as a substance (element, chemical or a compound of synthetic or molecular nature or biological compound(s)) that reduces the deleterious effects of radiation, when administered to living organisms, usually prior to irradiation (prophylactic administration). Agents that are effective during or immediately after radiation exposure are termed mitigators, while radiorecovery agents are those that help in recuperation and augment recovery once the living organism has been exposed to radiation. Most agents available today fall in the former category. Since the damage resulting due to ionizing radiation in most cases cannot be totally reversed, the currently available radiorecovery agents exhibit limited efficacy. There is immense interest now to develop all the three categories of agents so that radiation damage can be treated holistically. Collectively, these agents are classified as radiation countermeasure agents (Arora *et al.*, 2008a).

It is highly probable that such life-saving radiation countermeasure agents would form an integral of the armamentarium available at the hands of emergency responders and the military personnel for use during a radiation emergency. Radiation exposures to humans are classified into two categories: planned and unplanned exposures. Most of the radio-modifying agents are likely to find use during planned exposures e.g. rescue missions during accidents, radiotherapy of cancer patients and short and long-term space missions. Consequently, attempts have been made to design and develop drugs that can protect living organisms against the deleterious effects of ionizing radiation-mediated damage and also help in recovery under post-irradiation conditions. Drugs that can protect mammalian system against lethal radiation were developed by several developed countries, including the USA, France, Germany, China and USSR, mainly during the 'cold war' years. Among the various drugs developed, the sulfhydryls were reported to be the most remarkable (Bump and Malaker, 1998; Nair *et al.*, 2001; Coleman *et al.*, 2003; Arora, 2008; Arora *et al.*, 2010b).

5. The Emergence of Amifostine (WR-2721)

The United States Army conducted a number of trials at its famous Walter Reed Army Research Institute in Maryland, USA over the last fifty years with the aim to develop efficient and safe radioprotectors. Amifostine was part of a U.S. Army classified research project to identify agents that could be used for protection of military personnel against radiation exposure in the event of nuclear warfare. More than 5000 compounds were synthesized and screened at experimental level at the Walter Reed Army Institute of Research till 1980's (Brown, 1967; Sweeney, 1979; Giambarresi and Jacobs, 1987; Weiss and Simic, 1988). Outcome of research was the discovery of Amifostine (WR-2721 or ethiophos; Fig. 3), a compound known to possess radioprotective properties and was introduced into clinical trials particularly to ameliorate damage caused by radiotherapy and anticancer drugs (Yuhas and Phillips, 1982; Spencer and Gao, 1995). It was also used by US astronauts, on their first visit to the moon to protect them against radiation.

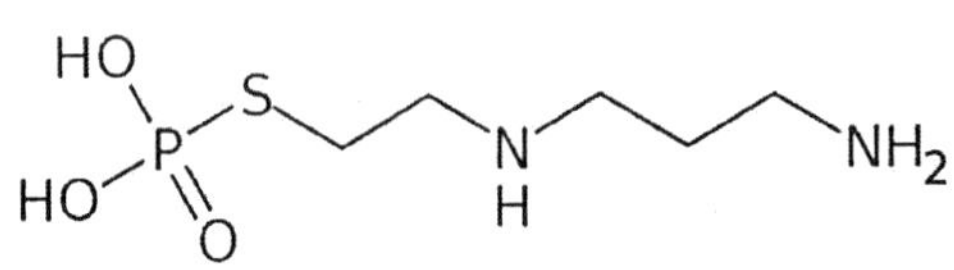

Fig. 3 : Amifostine: Often considered the 'Gold standard' in radioprotection. However, its acceptance in field conditions is fraught with a number of limitations.

Amifostine (WR-2721) has been the most famous radioprotector till date. Amifostine made its way into the US Army soldier kit, for protection against radiation-induced lethality but did not find acceptance due to the side effects viz., induction of nausea and vomiting, poor bioavailablity *via* oral-route, as it gets hydrolysed in the acidic environment of the stomach, is non-effective in post-administration scenario and exhibits a narrow therapeutic window (Mc Donough *et al.*, 1992; Tannehill and Mehta, 1996; Capizzi and Oster, 1995; Mcubcje *et al.*, 2003). Several phosphorothioates and thiosulphonates like WR-159243, WR-3302, WR-1551 and WR-2926 have also exhibited radioprotective properties. However, most of these compounds are not in use due to associated toxicity (Grdina *et al.*, 1991; Capizzi and Oster, 1995).

Amifostine (WR-2721) stands apart as a 'Gold standard'. Amifostine is a prodrug and is metabolized by the enzyme alkaline phosphatase to an active sulfhydryl compound (WR-1065) capable of scavenging radiation-generated free radicals and preventing cell damage.

During development of amifostine, animal studies were conducted to evaluate the radioprotective effects in various treatment protocols. A majority of these studies found that the use of amifostine protected normal tissues from toxicity to varying degrees. Similarly, the erstwhile USSR Army and several other European nations put in a lot of effort to develop such drugs. In recent years, scientists of the All-Russian Research Institute of Experimental Physics and their colleagues from the Lomonosov Moscow State University and St. Petersburg Chemistry and Pharmaceutical Academy are endeavoring to develop radioprotectors that are effective, harmless and inexpensive.

Amifostine is still used in the clinic; however, the serious side effects associated with its use have hindered its use on a large-scale.

6. The Global Scenario: Emerging trends

Research on the development of radioprotectors was initiated almost 50 years ago (Livesey and Reed, 1987). Patt *et al.* (1949) was amongst the pioneers who reported the protective role of Cysteine (a sulphur containing amino acid) against lethal dose of X-radiation.

A number of sulphydryl compounds have been screened since then. Research in the 80's focused on the synthetic drug screening using animal model systems, and gradually shifted to understanding the mechanism of action (Wasserman, 1994; Maisin, 1998). Several radioprotective compounds were also developed in the former Soviet Union and other East-European countries, of course under a cloud of extreme secrecy.

One of the radioprotective compounds acting *via* receptor includes eicosanoids (inflammatory mediators derived from arachidonic acid) and particularly leukotriene-C_4 has rendered increased survival of hemopoietic stem cells exposed to gamma-radiation. However, the associated behavioral toxicity and deleterious physiologic effects at even very low doses limits its clinical usage.

Combined modalities have also been tested for radioprotective efficacy. A combination modality of pre-administration of phosphorothioates with post-administration of immunomodulatory agents (having higher margin of safety, but lower efficacy when administered alone) has been reported to synergistically reduce the radiation damage (Weiss *et al.*, 1990; Neta, 1997). During the course of evolution, the natural compounds including minerals and vitamins (A, C, E) evolved as a part of the body's intrinsic antioxidant defense system, particularly to render protection to the cells against ROS (which are the result of normal metabolism) and have also been shown to promote recovery of wounds and burn injuries. The toxicity, metabolism, and pharmacological mode of action of these compounds have been worked out to a large extent in humans (Foye, 1969; Weiss and Kumar, 1988; Konopacka, 1996).

In the last decade, tremendous advances in biotechnology and a better understanding of the cellular effects of radiation has led to the development of several radioprotective agents (Nair *et al.*, 2001). The sequencing of human genome has generated tremendous information required for studying the multi-gene expession analysis and has initiated an era of ligand targeted drug discovery. Several products using recombinant DNA technology (interferon, interleukin, GM-CSF) have also been produced and tested for their radioprotective efficacy, particularly for the treatment of radiation injury victims (Browne *et al.*, 1990; MacVittie *et al.*, 1996).

Recent advances in stem cell technology, particularly for replenishing the hemopoietic, gastro-intestinal and central nervous system would find practical application in coming years. Non-specific immunomodulators that are able to induce cells to release a variety of peptides (cytokines) that act specifically on hemopoietic progenitor and stem cells to stimulate their growth and differentiation into mature functionally active cells are receiving attention (Neta, 1997; Venkatachalam and Chattopadhyay, 2005).

In recent years, multi-gene expression analysis has initiated ligand-based drug discoveries, while recombinant DNA technology based products like interferon, interleukin, Genetically Modified Colony Stimulating Factor (GM-CSF) have also been produced and tested for their radioprotective efficacy in the recent past. Recent advances in stem cell technology and subsequent development of non-specific immunomodulators, hemopoietic growth factors and cytokines like interleukin-7, interleukin-11, granulocyte-colony stimulating factor (G-CSF), granulocyte, macrophage-colony stimulating factor (GM-CSF), anti-apoptotic cytokine combinations and epithelial cell-specific growth factors like keratinocyte growth factor have also been found to mitigate radiation-induced damage though the augmented recovery of stem cells and their precursors was found to be limited.

The biggest problems associated with most synthetic radioprotectors (molecular drugs) are their high toxicity, low therapeutic index, short window of protection and lack of holistic mode of action. World-over a large number of chemical and biological compounds have been evaluated for radioprotection (Weiss *et al.*, 1990; Weiss and Landauer, 2003; Venkatachalam and Chattopadhyay, 2005). Several drugs like antioxidants, cytoprotective agents, angiotensin I-converting enzyme (ACE) inhibitors, metalloelements, non-steroidal anti-inflammatory drugs (NSAIDS), immunomodulators, sulphydryl compounds lipopolysaccharides and prostaglandins and DNA binding ligands have been evaluated for attenuating radiation-mediated injuries (Bump and Malaker, 1998; Lee and Stupans, 2002). Combination of biological and chemical agents has also been tried with some degree of success.

A number of hemopoietic growth factors and cytokines like interleukin, colony stimulating factors, stem cell factor, and anti-apoptotic cytokine combinations has been evaluated for radioprotection and radiorecovery (Singh and Yadav, 2005; Thierry *et al.*, 2005).

The beneficial aspects of Nanotechnology have been exploited in recent years to improve the efficacy of radioprotective formulations. The improvements in terms of efficacy enhancement have been substantial in some cases e.g, podophyllum and silymarin (Arora *et al.*, 2005b). More studies are needed to validate this technology so that the benefits can be reaped in coming years.

In spite of the intensive research efforts, the fact remains that till date there is not a single radioprotective agent available for human application that meets all the ideal requisites i.e., possesses no toxicity and offers long-term protection (Davidson *et al.*, 1980; Maisin, 1998; Nair *et al.*, 2001; Weiss and Landeur, 2003; Arora *et al.*, 2005a,b). In view of this, there has been a paradigm shift towards herbal drugs which are being viewed as potential radiation countermeasure agents (Weiss *et al.*, 1990; Arora and Goel, 2000; Goel *et al.*, 2000a,b,c,d; Goel *et al.*, 2001a,b; Weiss and Landauer, 2003; Arora and Sharma, 2004; Venkatachalam and Chattopadhyay, 2005; Sharma and Arora, 2006; Arora, 2006; Arora, 2007a,b).

7. Harnessing Indigenous Bioresource for Developing Radiation Countermeasure Agents

A plethora of plants and natual plant products have been evaluated in recent years and shown to possess radioprotective properties. The traditional wisdom of Indian, Greek, Korean, Iranian, Russian, Chinese, Japanese, Siddha, Sri Lankan, Indonesian, European, Tibetean and Unani systems of medicine recommends the usage of a combination of plants for treating diseases since synergistic effects can be obtained and some of the toxic effects generated by active constituents of one plant are neutralized by the presence of several other phytoconstituents. Medicinal plants contain a variety of compounds that act in a holistic manner and this is in particular needed for achieving radioprotection. By prospecting for plants possessing diverse biological activities, including radiation protection properties, new drugs can be designed. Several workers have worked out these properties in edible plants (useful as nutritional supplements and completely safe, but therapeutically less potent), low or high-altitude plants with reported medicinal values (quicker in action and therapeutically safer as compared to synthetic compounds). Probably, the combined regimen of both these categories of plants, with focused protocol for tackling different syndromes induced by radiation, could give a practical solution for radiation sickness. The organ-specific protection (specifically cerebroprotection, neuroprotection, renal protection), usage in cancer therapy, antioxidant potential, anti-inflammatory activities, metal chelation potential, free radical scavenging activity, DNA protectant property, anti-lipid protection properties, adaptogenic properties, usage against burns, wounds and skin allergies, immnuostimulatory, anti-ulcerogenic and tonifying potential of the plants are some of the properties that are generally investigated for radioprotective screening (Arora *et al.*, 2005a).

Prolonged effects of ionizing radiation are mainly elicited in digestive system, hematopoietic system, hepatic system, nervous system and reproductive system. Protecting only one single target system cannot render effective radioprotection. Radioprotection is a multi-faceted phenomenon and drugs that can protect multiple target organs are definitely effective. Herbal drugs contain several bioactive molecules and these can target and protect several organ systems simultaneously. The consolidated effect of the different biomolecules is manifested as whole-body radioprotection. Systematic studies have been carried out in recent years and a number of plants have been shown to render effective radioprotection at whole-body level. However, if some of these drugs in combination protect and target different organs simultaneously, the protective effect is increased manifold.

8. Herbal Radioprotectors and their Unique Mode of Action

Herbal radioprotectors act *via* multifarious mechanisms including metal chelation, free radical scavenging, immunomodulation, hemopoietic stimulation, enhancement of antioxidant defense, anti-inflammatory action, anti-microbial, cell proliferative, wound and skin healing activity, cyto-protection, DNA protection and maintenance of its repair processes and protection to various organ systems (mainly renal, gastro-intestinal, neural, hepatic and reproductive systems) [Arora *et al.*, 2005a,b]. At tissue/cellular level, some of the mechanisms include maintenance of membrane integrity (anti-lipid peroxidation activity), protein protection (chiefly antioxidant enzymes), mitochondrial integrity (membrane potential and energy status of cell), induction chromosomal DNA compaction (making it resistant against damage), induction of synthesis of pro-apoptotic factors and subsequently decreased levels of apoptotic inducers (decrement of radiation-induced apoptosis), and briefly 'switching on' of the cell-survival pathways by removal of unstable oxidant species from the system (Arora *et al.*, 2005a; Chawla *et al.*, 2005; Puri *et al.*, 2005; Sinha *et al.*, 2005). The holistic approach of herbal extracts makes them indispensable for targeting multifarious targets in a single dosage due to the uniform distribution of the active constituents from the antioxidant cocktail of herbal extracts.

9. The Cold Arid Zone High Altitude Plants as Potential Radiation Countermeasure Agents

The Institute of Nuclear Medicine and Allied Sciences (INMAS), DRDO, New Delhi has done pioneering work on herbal radioprotectors, particularly on the high-altitude cold arid plants under the aegis of several major R&D programmes. The leads obtained from previous R&D programmes were further pursued as a part of the 'CHARAK' (Herbs for our Armed Forces) Programme-a multi-laboratory DRDO program, with the ultimate aim of developing herbal radioprotective formulations for human applications, which would find use in both the civil and defence sector. An overview of studies carried out on the cold-arid high-altitude region bioresource is presented.

The high-altitude cold-arid region medicinal plants are very rich in a variety of phytochemicals and can be an important source of several drugs. Since these plants grow in extremely harsh environment, where the vagaries of nature provide them with an opportunity to evolve new phytomolecules, they can be useful as a source of new chemical entities.

A number of such plants from the cold-arid high-altitude region were screened for their radioprotective ability. More than 50 plants of the region were screened. Some of the plants screened included: *Aconitum heterophyllum, Arnebia euchroma, Carum carvii, Dactylorrhiza hatagirea, Ephedra gerardiana, Hypericum perforatum, Inula racemosa, Plantago, Rheum sp., Rubia cordifolia, Thymus vulgaris, Tribulus terrestris etc.* Out of the plants screened, *Hippophae rhamnoides, Podophyllum hexandrum* and *Rhodiola imbricata* gave very promising results and have been discussed in detail in the ensuing section (Fig. 4).

Seabuckthorn: The Golden Berry Bush

Hippophae rhamnoides Linn. (Fig. 5) is a plant of the cold arid high-altitude region and belongs to the family Elaeagnaceae. It is commonly known as seabuckthorn or Chuma, Ames, Dhurchuk or Milech, Tarru. The plant is a spiny shrub or small tree, about 2-4 m in height, berries are fleshy and are an important source of antioxidants. The plant is hard, drought and usually cold-tolerant plant and shoots are modified into spines. The leaves are lanceolate linear, obtuse, with peltate and stellate scales on lower surface. Male flowers are clustered at the base of shoot; petals free suborbicular, while female flowers have pedicellate inclusions (2-4), petals one 3-4 mm long, united utricles subglobose, succulent, red and orange coloured. Seeds are solitary, unequally bilobed, light, black and very hard. The plant is widely distributed throughout the temperate zones of Asia and Europe. Distribution ranges from the Himalayan region, India, Nepal, Bhutan, Pakistan, Afghanistan to China, Mongolia, Russia, Kazhakyzstan, Hungary, Romania, Switzerland, Germany, France and Britain to Scandinavian countries, including Finland, Sweden and Norway. In India, it is found in the North-West Himalayas.

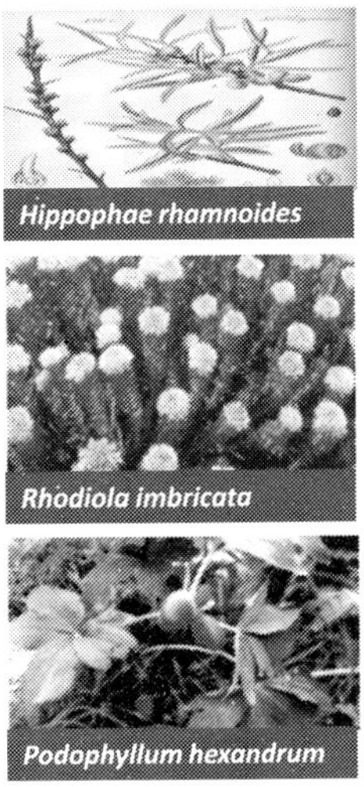

Life saving against Lethal Dose of Ionizing Radiation

Very useful in sub-lethal radiation doses

High degree of Efficacy and Safety

Radioprotective phytoconstituents identified

Protection to Multiple Target Organs/Systems

- *Hemopoietic system*
- *Gastrointestinal system*
- *Immune system*
- *Reproductive system*
- *Central nervous system*
- *Cutaneous system*

Mode of action at whole body, organ, tissue, cell and molecular level elucidated

Fig. 4 : The promising radioprotective plants of the cold arid high-altitude region.

Fig. 5 : Seabuckthorn: The Edible Golden Berry Bush has immense potential as a radioprotector.

The main phyto constituents include: carotene, and soluble sugars. The berries are rich in vitamin C content and Calcium, Chromium, Cobalt, Iron, Magnesium, Manganese, Molybednum, Potassium, Sodium, Strontium, and Zinc have been reported to be present. Malic and quinic acid are the major (>90%) organic acids present in fruit juice of species of different origin. The volatiles responsible for aroma include 3-methyl butanol, butyl pentanoate, 2-methyl propyl 3-methyl butanoate, ethyl 2-methyl butanoate and ethyl hexanoate. Quercetin, isorhamnetin, rhamnetin, steroids, terpenoids, alkaloids, fatty oil etc are also reported to be present. *H.rhamnoides* has been extensively used for the treatment of circulatory disorders, vaginal mucositis, oral mucositis, wound healing, cutaneous eruptions, lung complaints, stomach malfunctioning, duodenal ulcers, cervical erosion, scalds, skin ulcers, hepatic injury, neoplasia and gastric cancers. It is also useful for treating skin ulcers, gastric disorders/cancers, mucositis, wounds and other high-altitude disorders. It has been shown to effectively protect mammals against lethal ionizing radiation and acts *via* multifarious ways, including free radical scavenging, immune modulation, boosting antioxidant defence etc. *H.rhamnoides* protects the hemopoietic system, gastrointestinal system and central nervous system against the deleterious effects of radiation and also mitigates gamma-radiation induced genotoxicity. At molecular level it has been shown that a large number of genes and proteins are modulated, thereby rendering radioprotection.

Recently, a preparation of *Hippophae viz.*, RH-3 in a dose dependent fashion has been found to interact with plasmid DNA (*p*UC18) resulting in reduced mobility of supercoiled form (Sureshbabu *et al.*, 2008). RH-3 at higher concentrations (> 0.4 mg/ml) almost completely prevented the migration of supercoiled form without interfering with mobility of open circular form indicating its ability to selectively interact with supercoiled form. Studies done with supercoiled or open circular form also revealed the binding specificity of RH-3 for supercoiled form of plasmid. Both inhibited radiation induced strand breaks and DNA interaction by RH-3 were found to be dependent upon pH and the order of efficacy was found to be acidic pH> neutral pH > alkaline pH. RH-3 in a dose-dependent manner inhibited radiation-induced frank single, double strand breaks as well as endonuclease IV detectable abasic sites (clusters).

Tiwari *et al.* (2009) have reported that a single prophylactic, intra-peritoneal dose of the leaf extract of sea buckthorn (*H. rhamnoides* L.)

(SBL-1) protected > 90 percent population of whole body ^{60}Co-gamma-irradiated (10 Gy) mice, against 0 percent protection in irradiated (10 Gy) control mice population. They examined the antioxidant potential of SBL-1 and its influence on spontaneous and radiation induced mutation and gene-conversion. The biochemical assays showed that under *in vitro* conditions, SBL-1 scavenged superoxide radicals. SBL-1 also scavenged hydroxyl radicals generated by Fenton reaction as well as by ^{60}Co-gamma-rays. Under *ex vivo* conditions, SBL-1 countered the radiation-induced haemolysis of mice RBCs as well as lipid peroxidation in mice liver cells. SBL-1, by itself was found to be non-mutagenic and non-recombinogenic when tested in diploid strain D7 of *Saccharomyces cerevisiae*. In comparison to untreated-irradiated controls, SBL-1 treatment prior to ^{60}Co-gamma-irradiation countered the radiation-induced cell kill and mutagenesis at all radiation doses. This study suggested that SBL-1 offered radioprotection by antioxidant action and by promoting error-free post-irradiation repair pathways.

Tiwari and Bala (2011) have recently investigated effects of the preparation from leaves of *H. rhamnoides* (SBL-1), on modification of immunosupression and inflammation in whole body irradiated (10 Gy) mice. One time treatment with SBL-1 before irradiation prevented the radiation induced (i) decrease immunoglobulin G. (ii) early release of high mobility group box 1 (HMGB1) protein (iii) increase in tumour necrosis factor-α, myeloperoxidase activity and lipid peroxidation, (iv) liver haemorrhage, stomach enlargement, spleen shrinkage, intestinal oedema and hair fall, (v) decrease in thoils and ferric reducing ability of plasma. HMGB1, a known therapeutic target, has role in inflammatory diseases, tissue repair and immunomodulation. This study suggested that SBL-1 countered radiation pathologies by modifying HMGB1 regulated inflammatory pathway and restoring adaptive immune response.

The radioprotective effects of *H. rhamnoides* leaf extract, designated SBL-1, was investigated in Swiss Albino strain 'A' mice. against 100 percent mortality in whole-body irradiated (^{60}Co-gamma-rays, 10 Gy) control. A single dose of the SBL-1 rendered >90 percent survivors when administered 30 minutes before irradiation and 90 percent to 80 percent survivors when administered 1 to 4 hours before irradiation. SBL-1 activated proliferation of hemopoietic stem cells countered a radiation-induced decrease in total thiols and an increase in free radicals in plasma and liver, inhibited lipid peroxidation and normalized the liver alkaline phosphatase activity. This study demonstrated high radioprotective potential of *H. rhamnoides* leaves, which are easy to obtain and are inexpensive.

9.2 Himalayan Mayapple: The Ancient Divine Drug

In Ayurveda, the plant Himalayan Mayapple (*Podophyllum hexandrum* Royle; Fig. 6) has been mentioned as 'Aindri'- a Divine drug (Singh and Shah, 1994). *P. hexandrum* has been used as a cure for allergic and inflammatory conditions of the skin, biliary fever, burning sensation, cold, constipation, cancer of brain, bladder and lung, erysipelas, Hodgkin's disease, insect bites, mental disorders, monocytoid leukemia, non-Hodgkin's lymphoma, rheumatism, septic wounds plague and venereal warts since ages (Chatterjee and Pakrashi, 1995; Singh and Shah, 1994; Beutner and Von Krogh, 1990; Blasko and Cordell, 1988; Wong *et al.*, 2000). The radioprotective properties of *P. hexandrum* has been studied in great detail in recent years. The plant Podophyllum belongs to the family Berberidaceae and is commonly known as: Bankakri, Papra, Himalayan mayapple, Banwangam, Venival Patvel. *P. hexandrum* is found at altitudes above >2500 m, in the Himalayan ranges of India, including Sikkim, Jammu and Kashmir, Himachal Pradesh and Uttaranchal. Two other species *viz., P. montanum* and *P. peltatum* are found in Western China and North America respectively. The plant contains Picropodophyllotoxin, Podophyllotoxin (Fig. 7), Epi-Podophyllotoxin, Podohyllotoxin-b-D-galactopyranoside, 4'-Demethyl podophyllotoxin, Quercetin-3-O-b-D-galactopyranoside, Kaempferol-3-O-glucoside, Deoxypodophyllotoxin and Podophyllin.

Fig. 6 : The Himalayan Mayapple and its rhizome- the source of radioprotective biomolecules.

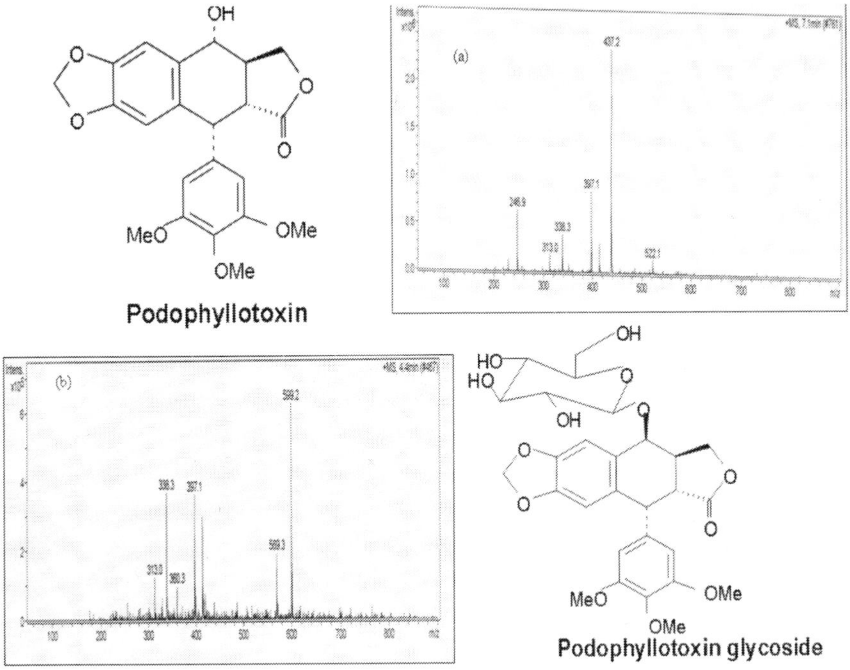

Fig. 7. The chemical structure and MS-MS profile of podophyllotoxin and its glycoside.

P. hexandrum is the source of podophyllotoxin and other aryltetralin lignans and several other bioactive constituents (Foster, 1993; Singh and Shah, 1994; Bedir *et al.*, 2002) several of which are used in anticancer therapy. Podophyllotoxin finds use as a precursor for the semi-synthetic anticancer drugs teniposide and itoposide and other anti-arthritic drugs (Fig. 8). These drugs and other derivatives, based on podophyllotoxin, find use for the treatment of leukemias, lung and testicular cancers, dermatological disorders like warts, rheumatoid arthritis, psoriasis and malaria (Leander and Rosen, 1988; Lerndal and Svensson, 2000).

Ionizing radiation has the capability of inducing severe of oxidative stress in living systems and causing mortality in mammals. In a 30-day survival study using Strain 'A' Swiss Albino mice, our group showed that the crude extract of *P. hexandrum* possessed the ability to save more than 80 percent animals with administration of a single dose two hours prior to 10 Gy exposure (Arora *et al.*, 2005a). Subsequent studies established that using *P. hexandrum* extract, a dose reduction factor (DRF; LD_{50}: treated group/untreated group) of 1.33 was achievable. In order to unravel the mode of action, the radiation-induced multi-organ dysfunction needs to be investigated at different levels of hierarchy of organization, and, therefore, our group investigated the organ-level protection rendered by *P. hexandrum* in greater detail. Podophyllum has been shown to protect various organs and systems, including the hemopoietic system, gastrointestinal system, reproductive system and central nervous system (Arora *et al.*, 2005a,b; Arora *et al.*, 2006; 2007a,b; Gupta *et al.*, 2008; Lata *et al.*, 2007; Shukla *et al.*, 2006; Sankhwar *et al.*, 2009; Gupta *et al.*, 2010). Studies by our group have shown that one of the mechanisms by which *P. hexandrum* renders radioprotection is modulation of radiation-induced oxidative stress. Podophyllotoxin and other aryltetralin lignans have been shown to play a crucial role in radioprotection. The mechanism of action has also been worked out at cellular and molecular level (Kumar *et al.*, 2007, 2008; Arora *et al.*, 2010a; Malhotra *et al.*, 2010).

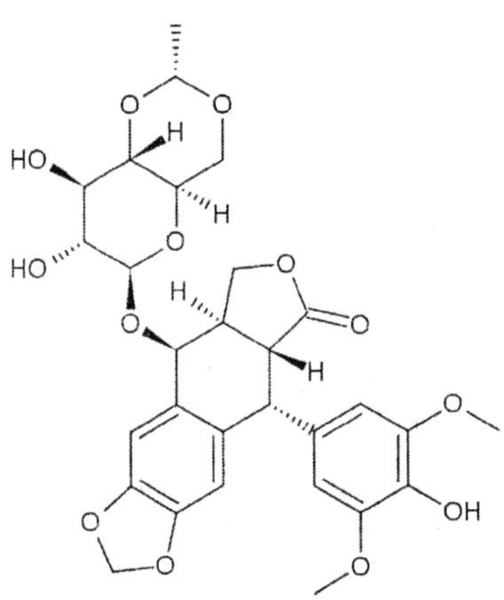

Etoposide

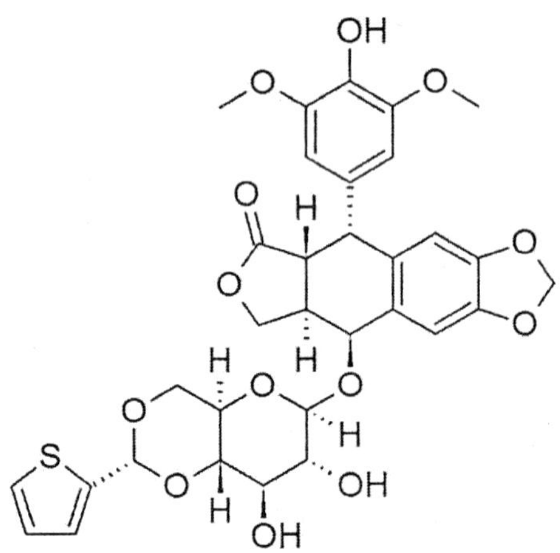

Teniposide

Fig. 8 : Etoposide and Teniposide: Two anticancer drugs derived from podophyllotoxin, which inhibit the activity of topoisomerase II.

The radioprotective efficacy of *P. hexandrum* fraction (REC-2006) in hepatoma cell lines having different p53 status has been reported in recent years. Higher radioresistance was observed in the HepG2 (p53(++)) cell line in comparison to the Hep3B (p53(-)) cell line, indicating a plausible role of p53 in radioresistance. REC-2006 exhibited nearly twice the survival in p53-expressing HepG2 cells compared with p53-negative Hep3B cells. REC-2006 treatment alone induced p53 expression as compared with untreated controls. However, REC-2006 reduced p53 expression when treated 2 hours before irradiation, as compared to the irradiated HepG2 controls, indicating that REC-2006 modulates the expression of p53 to mitigate its apoptotic effect. Induction of p21 in the REC-2006 + radiation treatment group downregulated the expression of cyclin E and CDK2 leading to a delay in the G1 phase of HepG2 cells, which provided time for DNA repair or related processes. However, no significant difference in CDC2 expression in both cell lines suggested that G2 phase arrest might not be the only responsible factor for REC-2006-mediated radioprotection. Significant induction of PCNA and GADD45 expression in HepG2 cells suggested that REC-2006 increased the percentage survival of HepG2 cells by increasing the span of time as well as efficacy for repair processes. REC-2006 modulated the expression of p53 and thereby promoted cell cycle arrest in the G1 phase, encouraging cell proliferation and DNA repair and thus providing significantly higher protection against acute gamma-radiation in the HepG2 cell line.

Gupta *et al.* (2010) have recently attempted to explore the mechanism of action of *P. hexandrum* sub-fraction (G-001M), which exhibited >90 percent protection in lethally irradiated mice. G-001M was assessed for antioxidant characteristics by evaluating DPPH, superoxide and hydrogen peroxide radical formation, anti-lipid per oxidation, metal chelation and total flavonoid content. To affirm cytoprotective efficacy of G-001M, plasmid DNA protection, blood WBC counts, marker for lipid peroxidation (MDA) and antioxidant status (GSH) in mice splenocytes and thymocytes were studied. G-001M, having high amount of total phenolic contents, exhibited dose dependent inhibition in DPPH and superoxide radical formation. Hydrogen peroxide radical scavenging was higher than standards. With pre-treatment of G-001M, plasmid DNA was also maximally restored to supercoiled form. Radiation modulated MDA and GSH values in splenocytes and thymocytes of mice altered significantly after 24 hrs and at later intervals, values were close to the controls. Radiation mediated losses in WBC counts were significantly regained ($p<0.001$) in G-001M pre-treated irradiated mice. The above findings explicitly conveyed that G-001M has successfully minimized radiation inflicted free radicals generation and their multiplication. This activity of G-001M could be undoubtedly among one of the major modes of action in extending whole body survival in lethally irradiated mice.

Sankhwar *et al.* (2009) found that the sub-fraction (G-001M), prepared from *P. hexandrum,* rendered high degree of survival (>90%) at a dose of 6 mg kg(-1) body weight (intramuscular) in lethally irradiated mice. Therapeutic dose of G-001M, at about 20 times lower concentration than its LD (100), has revealed a DRF of 1.62. Comet assay studies in peripheral blood leukocytes have reflected that, treatment of G-001M before irradiation has significantly reduced DNA tail length ($P < 0.001$) and DNA damage score ($P < 0.001$), as compared to radiation-only group. Spleen cell counts in irradiated animals had declined drastically at the very first day of exposure, and the fall continued till the 5th day ($P < 0.001$). In the treated irradiated groups, there was a steep reduction in the counts initially, but this phase did not prolong. More than 60 percent decline in thymocytes of irradiated group animals was registered at 5 hrs of irradiation when compared with controls, and the fall progressed further downwards with the similar pace till 5th day of exposure ($P < 0.001$). At later intervals, thymus was found fully regressed. In G-001M pre-treated irradiated groups also, thymocytes decreased till the 5th day but thereafter rejuvenated and within 30 days of treatment the values were close to normal. Current studies have explicitly indicated that, G-001M in very small doses has not only rendered high survivability in lethally irradiated mice, but also protected their cellular DNA, besides supporting fast replenishment of the immune system.

The results obtained from *P. hexandrum* point out that the plant is indeed very promising for radioprotection.

9.3 The Cosmonant's Plant: The Genus Rhodiola

Another high-altitude cold region plant is *Rhodiola imbricata* Edgew (Fig. 9), which is a member of the Family Crassulaceae is an erect perennial herb with rose-scented massive rootstock and fleshy, succulent stem. It has pale yellow flowers, which grow in clusters. It is found at heights of >4000 m in the high altitude Himalayan Ladakh region and cold arid regions of Pakistan and Nepal.

The biologically active phyto-constituents present in different *Rhodiola* species vary greatly according to the species in question (Brown *et al.,* 2002). Rhodiola contains chemical groups like Tannins, Phenolic glycosides, Phenylpropanoids (rosavin, rosin, rosarin), Flavonoids (rhodiolin, rhodionin, tricin, acetylrodalgin, catechins), Proanthocyanins (Rhodioloside), Phenylethanol derivatives (tyrosol), Phenolic acids (chlorogenic, hydroxycinnamic, caffeic and gallic acid),

Monoterpenes (rosaridin, rosaridol), Triterpenes (beta-sitosterol, daucosterol). The plant has been in use in traditional systems of medicine since antiquity, where it has been used for treating a variety of ailments and also as a prophylactic and general tonic. The plant increases physical endurance, work productivity, longevity and is used to treat fatigue, asthma, fever, hemorrhage, depression, anemia, impotence, gastrointestinal ailments, infections and central nervous system disorders. It is used for restoring memory and as a general health tonic and is also useful in cold, cough and lung ailments. In recent years, the plant has been used in Russia, Mongolia, China, India, America, Kazakhyzstan and European countries. The therapeutic uses include application in asthenic conditions like sleep disturbances, decline in work performance, loss of appetite, irritability and hypertension. It has been a widely accepted medicinal plant possessing antidepressant, tonifying, stimulative, DNA repair enhancing, anticancer, antimutagenic, reactive oxygen scavenging, adaptogenic, antiageing, antioxidant, anti-inflammatory, cardioprotective and central nervous system protective properties (Arora *et al.*, 2008: Chawla *et al.*, 2010).

Fig. 9 : Rhodiola: The cosmonaut's plant has found application as an adaptogen and a potent radiation countermeasure agent.

In a 30-day survival study, *R. imbricata* was found to be maximally effective at 400mg/kg body weight administered 30 minutes prior to 10 Gy exposure in Strain 'A' Swiss albino mice model system, 1000 mg/kg b.wt. indicating the non-toxic behaviour of the extract

(Arora *et al.*, 2005c, Arora *et al.*, 2008). The reported therapeutic activities of *R. rosea*, a close species to *R. imbricata*, and the present study matched indicating that it is an excellent candidate for investigation of biological activity particularly related to radiation protection. Comparative analyses of the biological activities associated with the plant have also been performed. Simultaneously, screening of the novel compounds as well as the different reported categories of compounds from this plant species, presumably responsible for radioprotection is also continuing. Studies have revealed the ability of the extract to effectively tackle radiation-induced oxidative stress (Chawla *et al.*; 2010).

Recently, Chawla *et al.* (2010) have compared the pro-oxidant/antioxidant activities of three fractionated extracts of *R. imbricata*. The aqueous fraction exhibited significant pro-oxidant activity under metal ion-induced stress ± flux [transition metal (Fe/Cr) ± 0.25 kGy]. A decrease in the dielectric constant of the solvent system utilized for extraction, exhibited a significant negative correlation (-0.955) with mean protection potential of lipid against radiation flux. Suchan effect was visualized as a significant shift from pro-oxidant to antioxidant activity in methanolic fraction (dielectric constant = 33), as compared to aqueous fraction (dielectric constant = 80). Aqueous fraction was predominantly pro-oxidant at maximal contentrations, indicating its anticancer potential. The presence of transition metals modulates such a biphasic activity differentially invarious fractions, i.e., the conversion of Fe(III) or Cu (II) to Fe(II) or Cu(I), respectively, due to the presence of certain bioactive constituents (electron donation at lower concentrations), favors pro-oxidant activity. On the other hand, certain other bioactive constituents involved in metal ion chelation contributed to its radioprotective efficacy. The aquo-methanolic fraction exhibited (disparate properties), i.e., concentration-dependent cytotoxicity (up to 250 μg/ml) and cytoprotection at 1000 μg/ml. *R. imbricata*, in general, exhibited a significant solvent-dependant variation in radioprotective efficacy. The authors concluded that solvent extraction and dose are crucial in bioactivity modulation and *R. imbricata* holds immense promise to be developed as a potential prophylactic radiation countermeasure agent for use in nuclear and radiological emergencies.

The *in vitro* antioxidant activities of aqueous extract of *R. imbricata* rhizome have also been reported by Gupta *et al.* (2009). Rhodiola extract was evaluated for DPPH (1,1-diphenyl, 2-picryl hydrazyl) radical scavenging, superoxide anion radical scavenging, hydrogen peroxide scavenging, anti-lipid peroxidation assay, metal chelating activity, reducing power and total flavonoid content. *R. imbricata* aqueous extract (RIAE) exhibited more DPPH scavenging potential with IC_{50} value of 5.56 μg/ml as compared to standard butylated hydroxytoluene (BHT) (IC_{50} = 5.74 μg/ml). The extract also provided superoxide anion radical scavenging (IC_{50} = 4.68 vs 4.63 μg/ml) and anti-lipid per-oxidation activity (IC_{50} = 5.06 vs 4.91 μg/ml) and IC_{50} value was comparable to standard α-tocopherol. The RIAE showed a decrease in hydrogen peroxide generation with IC_{50} value of 3.25 μg/ml when compared with standard α-tocopherol (IC_{50} value=4.00 μg/ml). IC_{50} of the extract for chelating activity was 5.33 μg/ml which was lower than the positive standard quercetin (IC_{50} = 3.12 μg/ml) but higher than α-tocopherol (IC_{50} = 6.13 μg/ml). The reducing capability of extract increased in dose dependent manner. Flavonoid content of RIAE determined spectrophotometrically in terms of quercetin equivalent was 66.7μg/mg. The results obtained in the present study indicated that aqueous extract of root of *Rhodiola imbricata* is a potent source of natural antioxidants. This study renders indirect support to the radioprotective properties of *R. imbricata*.

10. Conclusion

The high altitude cold arid zone plants *viz., Hippophae rhamnoides, Podophyllum hexandrum* and *Rhodiola imbricata* have shown promising radioprotective efficacy and hold immense potential for development as effective radiation countermeasure agents. Studies carried out in lower animal models have yielded sufficiently convincing data regarding their efficacy and safety. There is now a need to carry out pre-clinical evaluation, followed by testing in clinical model systems. The use of higher (primate) model systems would greatly help in getting regulatory clearance for use of drugs derived from these plants in humans, mainly during nuclear/ radiological emergencies.

A major challenge today is the development of efficient radiation countermeasure agents that are stable, effective, safe and usable for practical application in the defence and civil sector which becomes available soon. The edible nature of *H. rhamnoides* and *R. imbricata* makes them lucrative candidates for use as radiation countermeasures. The need of the hour is to focus attention on this extremely important activity in a concerted manner synergizing the expertise and resources available so that the radiation countermeasure agents are made available to the first responders, rescue personnel, military and civilian population for use during nuclear emergencies. It is imperative for our country to achieve self-reliance as no country would willingly share the expertise in this due to strategic and tactical reasons. To safeguard our national interests in the changing world scenario, it is imperative that we strive to develop the radiation countermeasure agents indigenously.

Acknowledgements

The authors gratefully acknowledge research grants and support received from the Defence Research and Development Organization (DRDO), Ministry of Defence, Government of India for the radiation countermeasure work. The authors also thank Ms Savita for her help with typing.

References

Arora, R. (2006) Phytodefence Molecules for Defence Forces: Utility and Future Prospects. In: Gupta, M., Kumar, I.P. and Arora, R. (eds) *Herbs for Health: Application in Armed Forces*. Institute of Nuclear Medicine and Allied Sciences, Delhi, India, pp. 58-62.

Arora, R. (2007a) Botanicals as Potential Radioprotective and Radiorecovery Agents: Current Status and Emerging Directions for Clinical Applications. *Proceedings of the World Ayurveda Conference*, Pune, India,

Arora, R. (2007b) Biological Radioprotection: A Phytochemical Perspective. *Proceedings of the Continuing Education Programme on Phytochemistry*. Defence Institute of Physiology and Allied Sciences, Delhi, India, December 10-14, pp. 19-37.

Arora R. (ed) (2008) Herbal Radiomodulators: Applications in Medicine, Homeland Defence and Space. CABI Publishing, Wallingford, Oxon, UK. pp. 332.

Arora, R. and Goel, H.C. (2000) Herbal Radioprotectors. *Proceedings of the International Conference on Radiation Biology (Radiobiology 2000)* Organized under the aegis of Indian Society for Radiation Biology by Regional Cancer Centre (RCC), Thiruvananthapuram, India from 17-19 February, 2000, pp. 87.

Arora, R. and Sharma, R.K. (2004) Development of Herbal Agents for Biological Radiation Protection. *Proceedings of the Defence Research and Development Organization (DRDO) Life Sciences Scientist's Conference*, Defence Food Research Laboratory, Mysore, India.

Arora, R., D. Gupta, R. Chawla, R. Sagar, A. Sharma, R. Kumar, J. Prasad, S. Singh, N. Samanta, and R.K. Sharma. (2005a). Radioprotection by Plant Products: Present status and Future Prospects. *Phytotherapy Research* 19(1): 1-22.

Arora, R., Chawla, R., Sagar, R.K., Singh, S., Kumar, R., Sharma, A.K., Singh, S., Prasad, J., Puri, S.C., Gupta, M.L. and Sharma, R.K. (2005b) Low aryl-tetralin lignans containing *Podophyllum hexandrum* extract protects against radiation-mediated lethal damage. *Proceedings of the Conference: Nanobiotechnology: The Road Ahead,* National Botanical Research Institute, Lucknow, India, pp. 29-30.

Arora R., Chawla R., Prasad J., Singh S., Kumar R., Sharma A., Sagar, Singh S., Sharma R.K. (2005c) Radioprotective activity of *Rhodiola imbricata*-a high altitude medicinal plant. *Molecular and Cellular Biochemistry* 273: 209-223.

Arora R., Dhaker A.S., Adhikari M., Sharma J., Chawla R., Kumar R., Sharma A., Gupta D., Prasad J., Singh S., Sinha A.K., Puri S.C., Qazi G.N., Ahuja P.S. and Tripathi R.P. (2008a) Strengthening the armamentarium of Indian Defence Forces for Radiological/Nuclear Preparedness: Potential of Radiation Countermeasure Agents of Natural Origin. In: In Service of the Soldier: Life Sciences Perspective (Editor Manas K. Mandal; Associate Editors: K. Ramachandran and A. Gupta). (Macmillan Advanced Research Series), Macmillan, New Delhi, India, pp.72-96.

Arora R., Chawla R., Dhaker A.S., Adhikari M., Sharma J., Singh S., Gupta D., Kumar R., Sharma A.K., Puri S.C., Sharma R.K. and Tripathi R.P. (2010a) *Podophyllum hexandrum* as a Potential Botanical Supplement for the Medical Management of Nuclear and Radiological Emergencies. *Journal of Dietary Supplements* 7(1), 31-50.

Arora R., Chawla R., Marwah R., Kumar V., Goel R., Arora P., Jaiswal S. and Sharma R.K. (2010b) Medical radiation countermeasures for nuclear and radiological emergencies: Current status and future perspectives. *Journal of Pharmacy and Bioallied Sciences.* 3;202-212.

Bedir, E., I. Khan, and R.M. Moraes. (2002). Bioprospecting of podophyllotoxin, trends in new crops and new uses; Herbs, medicinal, and aromatics, ASHS Press, Alexandria, VA, USA.

Beutner, K.R. and G. Von Krogh. (1990). Current status of podophyllotoxin for the treatment of genital warts. *Semin Dermatol* 9: 148.

Blasko, G. and G.A. Cordell. (1988). Recent Developments in the Chemistry of Plant derived Anticancer Agents. In: Wagner, H., H. Hiroshi, and N.R. Farnsworth, *eds.*, Economic and Medicinal Plant Research, Academic Press, London, United Kingdom.

Brown P.E. (1967) Mechanism of action of aminothiol radioprotectors. *Nature* 213: 363-364.

Brown, R.P., P.L. Gerbarg, Z, Ramazanov. (2002). *Rhodiola rosea*: A phytomedicinal overview. *Herbal Gram* 56: 40-52.

Browne D., Pillai M.V., MacVittie T.J., Weiss J.F. (1990). Treatment of Radiation Injuries. New York: Plenum Press.

Bump, E.A. and K. Malaker. (1998). Radioprotectors: Chemical, Biological and Clinical Perspective. CRC Press, Boca Raton, Florida, 431p.

Capizzi R.L., Oster W. (1995). Protection of normal tissue from the cytotoxic effects of chemotherapy and radiation by amifostine clinical experience. *Eur J Cancer* 31A(Suppl1): S8-S13.

Chatterjee, A. and Pakrashi. (1995). The Treatise on Indian Medicinal Plants, Vol 4, New Delhi, pp. 325.

Chawla R., Jaiswal S., Kumar R., Arora R., Sharma R.K. (2010). Himalayan bioresource *Rhodiola imbricata*: A Promising Radiation Countermeasure Agent for Nuclear and Radiological Emergencies. *J Pharm Bioall Sci*, 2 (3): 213-219.

Chawla, R., R. Arora, R. Kumar, A. Sharma, J. Prasad, S. Singh, R. Sagar, P. Chaudhary, S. Shukla, G. Kaur, R.K. Sharma, S.C. Puri, K.L. Dhar, G. Handa, V. Kumar, Gupta, G.N. Qazi. (2005). Antioxidant activity of fractionated extracts of rhizomes of high-altitude *Podophyllum hexandrum*: Role in Radiation Protection. *Molecular and Cellular Biochemistry* 273, 193-208.

Coleman N.E., Blakely W.F., Fike J.R., Mac Vittie T.J., Metling N.F., Mitchell J.B., Moulder J.E., Preston J.R., Seed Tm, Stone H.B., Tofilon P.J., Wong R.S.L. (2003) Molecular and cellular biology of moderate-dose (1-10Gy) radiation and potential mechanisms of radiation protection: Report of a Workshop (Bethesda, Maryland, USA, December, 17-18, 2001) *Radiat Res.* 159: 812-834.

Davidson D.E., Grenan M.M., Sweeney T.R. (1980) Biological characteristics of some improved radioprotectors. In: Radiation sensitizers, their use in the clinical management of cancer (Brady LW, ed.). New York: Masson, pp. 309-320.

Foster, S. (1993). Medicinal plant conservation and genetic resources: examples from the temperate Northern hemisphere. *Acta Hort.* 330: 67-73.

Foye W.O. (1969) Radiation-protective agents in mammals. J Pharmaceut Sci 58: 283-300.

Giambarresi, L. and A.J. Jacobs. (1987). Radioprotectants. In: Conklin, J.J. and R.I. Walker, *eds.* Military Radiobiology, Academic Press, Inc, London, pp. 265-301.

Goel H.C., Arora R., Prasad J., Sharma A.K., Singh S., Mathew T.L., Chaurasia O.P., Singh B. (2000b) A Process for Preparation of a Radioprotective Herbal Extract-I. Indian Patent, Indian Patents Office, Delhi, India.

Goel H.C., Arora R., Prasad J., Sharma A.K., Singh S., Mathew T.L., Chaurasia O.P., Singh B. (2000c) A Process for Preparation of a Radioprotective Herbal Extract-II. Indian Patent, Indian Patents Office, Delhi, India.

Goel H.C., Arora R., Shobi V., Mathew T.L. (2000d) A Process for Preparation of a Behavioural Radioprotective Herbal Extract. Indian Patent, Indian Patents Office, Delhi, India.

Goel H.C., Arora R., Sinha A.K., Joshi B.P., Prasad J., Singh S., Prem Kumar I., Sagar R., Salin C.A., Agarwala P.K., Prakash H., Gupta D., Sharma A.K., Mittal A., Pathania V., Mathew T.L., Singh B. (2000a) A Process for Preparation of a Radioprotective Herbal Extract from *Hippophae rhamnoides*. Indian Patent, Indian Patents Office, Delhi, India.

Goel H.C., Kumar I.P., Samanta N., Arora R. (2001a) A Process for Preparation of an Anticancer and Radioprotective Herbal Extract. Indian Patent, Patent Office, New Delhi, India.

Grdina D.J., Wright B.J., Carnes B.A. (1991). Protection by WR-151327 against late-effect damage from fission-spectrum neutrons. *Radiat Res* 128: S124-S127.

Gupta M., Sankhwar S., Verma S., Devi M., Samanta N., Agrawala P.K., Kumar R., Singh P.K. (2008) Whole body protection ot lethally irradiated mice by oral administration of semipurified fraction of *Podophyllum hexandrum* and post irradiation treatment of *Picrorhiza kurroa*. Tokai *J Exp clin Med* 33(1), 6-12.

Gupta V., Lahiri S.S., Sultana S., Tulsawani R., Kumar R. (2009) In vitro evaluation of antioxidant and free radical scavenging activities of *Rhodiola imbricata*. *Journal of Complementary and Integrative Medicine* 6(1), 1-17.

Gupta M.L., Gupta V., Shukla S.K., Verma S., Sankhwar S., Dutta A., Suri K.A. (2010) Inhibition in radiation mediated cellular toxicity by minimizing free radical flux: one of the possible mechanisms of biological protection against lethal ionizing radiation by a sub- fraction of *Podophyllum hexandrum*. *Cell Mol Biol* (Noisy-le-grand). 11(56): 1341-1349.

Konopacka, M. (1996). Vitamins as radioprotectors of normal cells. Postepy Hig Med Dosw 50: 145-156.

Goel H.C., Arora R., Prasad J., Sharma A.K., Singh S., Mathew T.L., Chaurasia O.P., Singh B. (2001b) A Radioprotective Herbal Formulation. Indian Patent, Indian Patents Office, Delhi, India.

Kumar R., Singh P.K., Arora R., Sharma A.K., Sharma R.K. (2007) Modulation of apoptosis-inducing protein in mouse kidney by *Podophyllum hexandrum*. In: Recent Progress in Medicinal Plants.Editor: JN Govil, VK Singh and Rakesh Bhardwaj Studium Press, USA. pp. 221-236.

Kumar, R., Singh, P.K., Arora R., Chawla, R. and Sharma, R.K. (2008) Radioprotective activities of *Podophyllum hexandrum*: Current knowledge of the molecular mechanisms. Trees for Life 2009; 4: 1 (www.trfjounral.org/ article:php/20081015155-249393).

Lata M., Prasad J., Singh S., Kumar R., Singh L., Chaudhary, Arora R., Chawla R., Tyagi S., Soni N.L., Sagar R.K., Devi M., Sharma R.K., Puri S.C., Tripathi R.P. (2007) Whole body protection against ionizing radiation in mice by REC-2001: A semi-purified fraction of *Podophyllum hexandrum*. Phytomedicine (in press).

Leander, K. and B. Rosen. (1988). Medicinal used for podophyllotoxin. US patent 4,788, 216:

Lee, T.K. and I. Stupans. (2002). Radioprotection: the non-steroidal anti-inflammatory drugs (NSAIDs) and prostaglandins. *J. Pharm Pharmacol* 54: 1435-1445.

Lerndal, R. and B. Svensson. (2000). A clinical study of CPH 82 vs. methotrexate in early rheumatoid arthritis Rheumatology (Oxford) 39: 316.

Livesey J.C., Reed D.J. (1987) Chemical protection against ionizing radiation. *Adv Radiat Res* 13: 285-353.

MacVittie T.J., Weiss J.F., Browne D. (1996) Advances in the treatment of radiation injuries. Oxford: Pergamon.

Maisin J.R. (1998) Chemical radioprotection: past, present and future prospects. *Int J Rad Biol* 73: 443-450.

Malhotra P., Adhikari M., Sharma J., Chawla R., Dhaker A.S., Kumar R., Sharma A., Gupta D., Sharma R.K., Tripathi R.P., Arora R. (2010) Optimizing radioprotective efficacy of a fractionated extract of Podophyllum hexandrum for mitigation of radiation-induced free radical-mediated biological effects in the hemopoietic syndrome range. Advanced BioTech (*Journal of Advanced Biotechnology*) 10(5), 128.

Mc Donough J.H., Mele P.C., Franz C.G. (1992) Comparison of behavioural and radioprotective effects of WR-2721 and WR-3689. *Pharmacol Biochem Behav* 42: 33-243.

Mcubcje V., Van Bcuningen D., Sohns T., Fliedner T.M. (2003) Medical management principles for radiation accidents. *Mil. Med.* 168: 219-222.

Metller F.A., Voclz G.L. (2002) Current concepts: major radiation exposure: what to expect and how to respond. *N Engl J Med* 346: 1554-1661.

Milosevic Z., Kljajic R., Horsic E., Hasanbasic D., Herceg Z. (1988) Clinical and Pathomorphological changes in the gastrointestinal tract in swine after irradiation with high X-ray doses. Radiation Protection: Advances in Yugoslavia and Italy, Proceedings of the Italian-Yugoslav Symposium, Udine, Italy, ENEA, pp. 119-123.

Nair, C.K.K., D.K. Parida, and T. Nomura. (2001). Radioprotectors in Radiotherapy. *J. Radiation Research* 159: 812-834.

Neta R. (1997) Insights into the mechanisms of radiation damage and radioprotection through their modulation by cytokines. Environ Health Perspect 105(Suppl 6): 1463-1465.

Patt H.M., Tyree E.B., Straube R.L., Smith D.E. (1949) Cysteine protection against X-irradiation. *Science* 110: 213-214.

Prasad K.N. (1999) Handbook of Radiobiology. CRC Press, Boca Raton, Florida, pp.344.

Puri S.C., Nazir A., Gupta V.K., Negi J.S., Handa G., Dhar K.L., Qazi G.N., Arora R., Kumar R., Sharma A.K., Prem Kumar, Prasad J., Singh S., Chawla R., Sagar R.K., Sharma R.K. (2005) A process for preparation of a radioprotective formulation containing aryltetralin lignans. Patent application filed with The Controller of Patents, Indian Patents Office, Delhi, India through CSIR.

Sankhwar S., Gupta M.L., Gupta V., Verma S., Suri K.A., Devi M., Sharma P., Khan E.A., Alam M.S. (2009) *Podophyllum hexandrum*-mediated survival protection and restoration of other cellular injuries in lethally irradiated mice. Evid Based Complement Alternat Med. Jun 24. [Epub ahead of print]

Sharma, R.K. and Arora, R. (2006) *Herbal Drugs: A Twenty First Century Perspective.* Jaypee Brothers Medical Publishers Private Ltd., Delhi, India, pp. 688.

Sharma R.K. and Arora R. (2011) Fukushima, Japan: An apocalypse in the making? *Journal of Pharmacy and Bioalllied Sciences* 3(2): 315-316.

Shukla S.K., Chaudhary P., Prem Kumar I., Afrin F., Puri S.C., Qazi G.N., Sharma R.K. (2006) Cytotoxic and radioprotective effects of *Podophyllum hexandrum. Environmental Toxicology and Pharmacology* 22: 113-120.

Singh, J. and N.C. Shah. (1994). *Podophyllum*: A Review. *Curr Res Med Arom Plants* 16: 53-83.

Singh, V.K. and V.S. Yadav. (2005). Role of cytokines and growth factors in radioprotection. *Experimental and Molecular Pathology* 78: 156-169.

Sinha A.K., Swaroop A., Chawla R., Arora R., Kumar R., Sharma A.K., Prem Kumar, Prasad J., Singh S., Chawla R., Sagar R.K., Sharma R.K. (2005) A process for preparation of a herbal radioprotective extract from *Hippophae rhamnoides*. Patent being filed through Indian Patent Office, Delhi, India through CSIR.

Spencer C.M., Gao K.L. (1995) Amifostine: A review of its pharmacodyanamic and pharmacokinetic properties, and therapeutic as a radioprotector and cytotoxic chemoprotector. Drugs 50: 1001-1031.

Sureshbabu A.V.S., Barik R.K., Namita I., Prem Kumar I. (2008) Radioprotective properties of *Hippophae rhamnoides* (sea buckthorn) extract in vitro. *Int. J. Health Sci* (Qassim) 2(2): 45-62.

Sweeney, T.R. (1979). A Survey of Compounds from the Antiradiation Drug Development Program of the US Army Medical Research and Development Command, Walter Reed Army Institute of Research, Washington DC, USA.

Tannehill, S.P. and M.P. Mehta. (1996). Amifostine and radiation therapy: past, present, and future. *Semin Oncol* 23:69-77.

Thierry, D., J.M. Bertho, A. Chapel and P. Gourmelon. (2005). *British J. of Radiology (Supplement)* 27: 175-179.

Tiwari S., Arya A., Tyagi S., Prasad J., Singh S., Vats P., Kumar D., Jain S.K., Balam. (2009) Antioxidant, antimutagenic and radioprotective properties of sea buckthorn leaf (*Hippophae rhamnoides* L.) Zeitschrift fur Arznei- and Gewurzpflanzen 14(2), 83-89.

Tiwari S. and Bala M. (2011) Hippophae leaves prevent immunosuppression and inflammationin 60Co irradiatied mice. *Phytopaharmacology* 1(3): 36-48.

Venkatachalam, S.R. and S. Chattopadhyay. (2005). Natural radioprotective agents: An overview. *Curr. Org. Chem.* 9: 389-404.

Wasserman T.H. (1994) Radiotherapeutic studies with amifostine (Ethyol). *Semin Oncol* 21: 21-25.

Weiss J.F. and Kumar K.S. (1988) Antioxidant mechanisms in radiation injury and radioprotection. In: Cellular Antioxidant Defense Mechanisms, Vol II (Chow CK, ed). Boca Raton, FL: CRC Press, pp. 163-189.

Weiss J.F. and Simic M.G. (1988) Perspectives in radioprotection. *Phamacol Ther* 39: 1-414.

Weiss J.F., Kumar K.S., Walden T.L., Neta R., Landauer M.R., Clark E.P. (1990) Advances in radioprotection through the use of combined agent regimens. *Int. J Radiat Oncol Biol* 57: 709-722.

Weiss, J.F. and M.R. Landauer. (2003). Protection against ionizing radiation by antioxidant nutrients and phytochemicals. *Toxicology* 189: 1-20.

Wong, S.K., S.K. Tsui, S.Y. Kwan, X.L. SU, and R.C. Lin. (2000). Identification of characterization of *Podophyllum emodi* by API-LC/MS/MS. *J Mass Spectrom* 35: 1246-1251.

Yuhas J.M., Philips T.L. (1982) Pharmacokinetics and mechanisms of action of WR-2721 and other protective agents. Radioprotector and Anticarcinogenesis. New York: Academic Press, pp. 639-653.

□□□

About the Editors

Dr. Ravi Bihari Srivastava, Ph.D, Outstanding Scientist and Director, Defence Institute of High Altitude Research (DIHAR) is heading the world's highest agro-animal laboratory at Leh Ladakh. He obtained his Post Graduation in Botany from Allahabad Univerisity and Ph.D from Kumaon University. Having joined DRDO in 1979, Dr. R.B. Srivastava was instrumental in establishing the Marine Biology Laboratory at Naval Materials Research Laboratory (NMRL) and was involved in several strategic research projects related to aerospace materials at Defence Materials & Stores Research & Development Establishment (DMSRDE), Kanpur. As Director of Defence Research Laboratory (DRL), Tezpur, Dr. R.B. Srivastava significantly contributed to the well being of the troops in the North-Eastern region of India through development of water purification systems for Iron and Arsenic removal, Herbal Antimalarials, Prophylactics against Black Fly bites, Microalgae based biodiesel production systems *etc.* During his tenure as Director, Directorate of Life Sciences (DLS), he prepared business plan and road map on futuristic research areas for the Life Science, group of laboratories in DRDO to be implemented through interdisciplinary collaborations between various laboratories. Having taken over as Director DIHAR since December 2010, Dr. R.B. Srivastava has been instrumental in devising research and development for augmentation of agro-animal technologies and optimal utilization of non-conventional energy resources in Ladakh region. He is an author of 18 book chapters, edited books, has more than 60 publications in peer reviewed journals and 09 patents to his credit. Dr. R.B. Srivastava has been honoured with several prestigious awards including the "DRDO Scientist of the Year Award" in recognition to his contributions for development of Bio-resistance coatings for War Equipments, evolving new concepts in Microbe-Metal Interactions and unveiling a new fungal species for biodegradation of polymers.

Dr W Selvamurthy, Ph.D. Distinguished Scientist is presently Chief Controller, Research & Development (Life Sciences (LS) and International Cooperation (IC)) in DRDO. He obtained his post-graduation in Human Physiology from Christian Medical College, Vellore in 1972, Ph.D from University of Delhi in 1982 and Doctorate of Science (D.Sc.) from Swami Vivekananda Yoga Anusandhana Samsthana, Bangalore in 2006. He has been awarded the Doctor of Science (D.Sc.) (Honoris Causa) in recognition of his eminent attainment in the field of science and his distinctive place in the scientific world by Fakir Mohan University, Balasore in 2008 and by Bharathiar University, Coimbatore in 2009. He has made significant research contributions for the benefit of the soldiers in the fields related to physiological acclimatization at high altitude and application of yoga for the armed forces. He has 12 books and over 170 research papers to his credit. He was the leader of the First Indo-Soviet Scientific Expedition to the Arctic for physiological experiments (1990-91). He is fellow of International Medical Sciences Academy (FIMSA) 2006, National Academy of Medical Sciences in 2008, President of Federation of Indian Physiological Societies in 1990 and Vice President of Indian Association of Biomedical Scientists. He is the recipient of several awards and honours in recognition of his outstanding research contributions *viz.* Technology Leadership Award, DRDO in 2010, Lifetime Achievement Award in the field of Clinical and Preventive Cardiology in the World (2006-2007), Dr. CL Baid Metha Oration in 2005, HH Loescheke Research Prize in Respiratory Physiology in 2002, National Citizen's Award in 2001, Scientist of the year Award' by DRDO in 1989 to name a few.